Natural Corrosion Inhibitors

Synthesis Lectures on Mechanical Engineering

Synthesis Lectures on Mechanical Engineering series publishes 60–150 page publications pertaining to this diverse discipline of mechanical engineering. The series presents Lectures written for an audience of researchers, industry engineers, undergraduate and graduate students.

Additional Synthesis series will be developed covering key areas within mechanical engineering.

Natural Corrosion Inhibitors
Shima Ghanavati Nasab, Mehdi Javaheran Yazd, Abolfazl Semnani, Homa Kahkesh, Navid Rabiee, Mohammad Rabiee, and Mojtaba Bagherzadeh
2019

Fractional Calculus with its Applications in Engineering and Technology
Yi Yang and Haiyan Henry Zhang
2018

Essential Engineering Thermodynamics: A Student's Guide
Yumin Zhang
2018

Engineering Dynamics
Cho W.S. To
2018

Solving Practical Engineering Problems in Engineering Mechanics: Dynamics
Sayavur I. Bakhtiyarov
2018

Solving Practical Engineering Mechanics Problems: Kinematics
Sayavur I. Bakhtiyarov
2018

C Programming and Numerical Analysis: An Introduction
Seiichi Nomura
2018

© Springer Nature Switzerland AG 2022

Reprint of original edition © Morgan & Claypool 2019

All rights reserved. No part of this publication may be reproduced, stored in a retrieval system, or transmitted in any form or by any means—electronic, mechanical, photocopy, recording, or any other except for brief quotations in printed reviews, without the prior permission of the publisher.

Natural Corrosion Inhibitors

Shima Ghanavati Nasab, Mehdi Javaheran Yazd, Abolfazl Semnani, Homa Kahkesh, Navid Rabiee, Mohammad Rabiee, and Mojtaba Bagherzadeh

ISBN: 978-3-031-79628-9   paperback
ISBN: 978-3-031-79629-6   ebook
ISBN: 978-3-031-79630-2   hardcover

DOI 10.1007/978-3-031-79629-6

A Publication in the Springer series
SYNTHESIS LECTURES ON MECHANICAL ENGINEERING

Lecture #18
Series ISSN
Print 2573-3168   Electronic 2573-3176

Natural Corrosion Inhibitors

Shima Ghanavati Nasab
University of Shahrekord, Shahrekord, Iran

Mehdi Javaheran Yazd
Islamic Azad University, Khomeinishahr, Iran

Abolfazl Semnani
University of Shahrekord, Shahrekord, Iran

Homa Kahkesh
Shahid Chamran University of Ahvaz, Ahvaz, Iran

Navid Rabiee
Sharif University of Technology, Tehran, Iran

Mohammad Rabiee
Amirkabir University of Technology, Tehran, Iran

Mojtaba Bagherzadeh
Sharif University of Technology, Tehran, Iran

SYNTHESIS LECTURES ON MECHANICAL ENGINEERING #18

ABSTRACT

Regarding the vast use of chemical components in different human activities, they are susceptible to corrosion because of contact with aggressive environments. Therefore, the use of inhibitors for the control of corrosion of chemical components in corrosive media is an urgent affair. Numerous investigations were carried out and are still being done to study the corrosion inhibition potential of organic compounds. Remarkable inhibition efficiency was achieved by all these compounds particularly some with N, S, and O atoms in their structure. Unfortunately, most of these compounds are not only expensive but also toxic to living beings. Hence, it is essential to mention the importance of natural inhibitors as eco-friendly, readily available, and renewable sources. The main goal of this book is to point out the influence of these components in terms of physical and, in some cases, chemical, processes on different components in aggressive media. Different methods of measuring corrosion, adsorption behavior, mechanism of inhibitors, and some related information is presented in this book. There is a lack of comprehensive and relevant books on this subject, so we decided to write this book in order to accumulate useful information about the influence of natural inhibitors on metals in corrosive areas and to make it accessible to researchers.

KEYWORDS

corrosion, corrosion inhibitors, adsorption, organic compounds, renewable sources

Contents

Preface

The most extensive used metal is iron. Different methods have been developed to control and prevent corrosion in many environments. Using inhibitors is one of the most pragmatic strategies to prolong corrosion progress which has become widespread recently. Different organic and inorganic barriers have been studied to control corrosion of metals. Most organic compounds include nucleophiles like nitrogen, sulfur, oxygen atoms, heterocyclic compounds, and -electrons which allow an adsorption on metal surface. These compounds can be adsorbed on metal surface and block active surface sites to decrease the corrosion rate. Although many organic compounds such as aliphatic amines, aromatic amines, aromatic aldehyde, aromatic acids, carbonyl compounds, phenol, thiosemicarbazide derivatives, hydrazine derivatives, amino acids, antibacterial drugs, Schiff bases, alkylimidazolium ionic liquids, organic dye, anionic surfactants, cationic surfactants, nonionic surfactants, pyridine derivatives, benzotriazole derivatives, triazoline derivatives, and tetrazole derivatives reveal good anticorrosive action, most of them are expensive, poisonous and enhance environmental concerns. These properties make them undesirable. Inorganic compounds like chromates, dichromates, and nitrates are also applied to control corrosion rate by acting as anodic inhibitors. Unfortunately, the biological toxicity of these inorganic, especially chromates and organophosphates are pointing toward their environmentally harmful properties. Although, many synthesized compounds manifest good corrosion inhibitor, most of them are highly toxic to both human beings and environment. These inhibitors can cause temporary or permanent damage to organ system like kidneys or liver, or to disorder a biochemical process or to disorder an enzyme system at some site in the body. These harmful factors lead natural inhibitors to develop. Recently most of attentions are focused on using natural inhibitors, for example plant seeds, leaves, plants aqueous extract, peels, fruits, pectin, lignin, oil extracts, plant alkaloids extracts, and etc. as the most important and favorable compounds for the environment and human. They have some pros over chemical inhibitors including being non-poisonous, cheap, environmentally friendly, renewable, and readily available which make them helpful. Inhibitive action of such barriers is attributed to some organic constituents of barriers, whose electronic structures resemble those of conventional organic corrosion inhibitors.

Regarding to the importance of using natural inhibitors in different industries such as petroleum industry and such like in order to control corrosion, their desirable properties, and lack of articles on this subject (however, there has been a short and letter review presented by *Mathur Gopalakrishnan sethuraman et al.* on the subject of natural products as corrosion inhibitor in corrosive media) we decided to provide such book to gather all related, essential, and comprehensive information about natural inhibitors and make it accessible to researchers. The main goal of creating this book is to review the impression of natural inhibitors worked by researchers on corrosion of metals and alloys in different aggressive areas. We classified inhibitors into some groups and mentioned the effect of them on corrosion rate, inhibition efficiency, and other important parameters. Different methods of measuring corrosion, adsorption behavior, inhibitors mechanism, and other related information is presented.

Shima Ghanavati Nasab, Mehdi Javaheran Yazd, Abolfazl Semnani, Homa Kahkesh, Navid Rabiee, Mohammad Rabiee, and Mojtaba Bagherzadeh
April 2019

CHAPTER 1

An Introduction to Natural Corrosion Inhibitors

Iron and its alloys are often used in the chemical and petrochemical industries. Acid solutions, especially hydrochloric acid solutions, are utilized for cleaning, pickling, descaling, and oil well acidizing. Hydrochloric acid solution is used more than others, because it is more economical, effective, and less trouble in comparison to mineral acids.

Thanks to the aggressiveness of acids, inhibitors are often applied to decrease the dissolution of metals. So, appraisement of the corrosion phenomenon on metals, especially different kinds of steel, and finding admirable methods are vital [1, 2].

Corrosion is the destruction of substances by chemical reaction with their environment. The word corrosion is also used in reference to the decay of plastic, concrete, and wood, but generally refers to metals.

The most extensively used metal is iron. Different methods have been developed to control and prevent corrosion in many environments [3]. Using inhibitors is one of the most pragmatic strategies to prolong corrosion progress which has become widespread recently [4].

Different organic and inorganic barriers have been studied to control the corrosion of metals [5]. Most organic compounds include nucleophiles like nitrogen, sulfur, oxygen atoms, heterocyclic compounds, and π-electrons which allow an adsorption on a metal surface [6]. These compounds can be adsorbed on a metal surface and block active surface sites to decrease the corrosion rate [7].

Although many organic compounds such as aliphatic amines [8, 9], aromatic amines [10], aromatic aldehyde [11], aromatic acids [12], carbonyl compounds [13], phenol [14], thiosemi-carbazide derivatives [15], hydrazine derivatives [16, 17], amino acids [18], antibacterial drugs [10, 19], Schiff bases [20–22], alkylimidazolium ionic liquids [23], organic dye [24], anionic surfactants [25–27], cationic surfactants [28, 29], nonionic surfactants [30], pyridine derivatives [31], benzotriazole derivatives [32], triazoline derivatives [33], and tetrazole derivatives [34] reveal good anticorrosive action, most of them are expensive and poisonous, and enhance environmental concerns. These properties make them undesirable [35].

Inorganic compounds like chromates [36], dichromates [37], and nitrates [38] are also applied to control corrosion rate by acting as anodic inhibitors. Unfortunately, the biological toxicity of these inorganic, especially chromates and organophosphates, are pointing toward their environmentally harmful properties [6].

Although many synthesized compounds manifest good corrosion inhibitors, most of them are highly toxic to both human beings and the environment. These inhibitors can cause temporary or permanent damage to organ systems like the kidneys or liver, to disorder a biochemical process, or to disorder an enzyme system at some site in the body. These harmful factors lead natural inhibitors to develop [7].

Recently, most attention has been focused on using natural inhibitors, for example plant seeds [39, 40], leaves [41, 42], plants aqueous extract, peels, fruits [5, 43–47], pectin [48], lignin [49], oil extracts [50, 51], plant alkaloids extracts [52, 53], etc., as the most important and favorable compounds for the environment and human. They have some pros over chemical inhibitors including being nonpoisonous, cheap, environmentally friendly, renewable, and readily available which make them helpful [54]. Inhibitive action of such barriers is attributed to some organic constituents of barriers, whose electronic structures resemble those of conventional organic corrosion inhibitors [55]. Regarding the importance of using natural inhibitors in different industries, such as the petroleum industry, and in order to control corrosion, their desirable properties, and due to a lack of articles on this subject (with the exception of an article by Raja and Sethuraman [56] on the subject of natural products as corrosion inhibitors in corrosive media), we decided to gather all related, essential, and comprehensive information about natural inhibitors and make it accessible to researchers.

The primary goal of writing this book was to review the impression of natural inhibitors worked on by researchers on the corrosion of metals and alloys in different aggressive areas. We classified inhibitors into groups and reviewed the effect of them on corrosion rate, inhibition efficiency (IE), and other important parameters. Different methods of measuring corrosion, adsorption behavior, inhibitors mechanism, and other related information are also presented.

CHAPTER 2

Corrosion Inhibitors: Fundamental Concepts

It is almost impossible to completely prevent corrosion, but it is possible to control it [57]. The use of inhibitors is one of the most applied techniques for preservation against corrosion and prevention of unanticipated metal dissolution and acid consumption, particularly in acid solutions [5].

A corrosion inhibitor is a substance that is efficient in very small amounts when added to a corrosive area to reduce the corrosion rate of the exposed metallic material [57]. These inhibitors are classified based on their action (as anodic, cathodic, and mixed inhibitors) and their mechanism of action (as hydrogen evolution, scavengers, vapor-phase, and adsorption inhibitors) [58].

Inhibitors decrease corrosion rate by: (1) increasing or reducing the cathodic or/and anodic reaction; (2) decreasing diffusion rate for reactants to the surface of the metal; and (3) reducing the electrical persistence of the metal surface [56]. Unfortunately, corrosion inhibitors are effective for a special metallic material in a certain area. Tiny changes in the composition of the solution or alloy can primarily vary the inhibition performance.

The election and the amount of inhibitors used depend on the kind of corrosive environment, its strength, kind of metal, favorable protection time, and expected temperature. One of the important factors in choosing inhibitors is the maximum temperature limit because some materials are vulnerable to thermic decomposition and they lose their inhibition effectiveness [57].

Various compounds are applied as corrosion inhibitors. Because the mechanism of how corrosion inhibitors work is usually unknown, experimental testing is still inevitable, despite some proposed models for predicting IE.

The scientific community and the industry do not completely notice the mechanism or the role of corrosion inhibitors and it is difficult or sometimes impossible to forecast whether or not a special compound will work [57].

Recently, because of environmental matters, researchers have been working on the concept of tiny harmful inhibitors to the environment (green inhibitors) to elude the poisonous effect of synthesized corrosion inhibitors [59].

Natural inhibitors have some advantages over chemical banners including being low cost, non-toxic, and eco-friendly, which has encouraged researchers to study and develop them; however, their IE is usually found to be very low [57]. Natural products, for example lupine extract [60], Gossipiumhirsutum L. extract [44], Murrayakoenigii leaves extract [61], Punicagranatum extract [62], Dacryodisedulisextract [55], damsissa (Ambrosia maritime, L.) extract [63], Oxandraasbeckii extract [64], Phyllanthusamarus extract [65], seed extract of Psidiumguajava [66], Launaea nudicaulis extract [67], and Artemisia pallens (Asteraceae) extract [68], etc., are an abundant source of naturally synthesized chemical compounds (e.g., amino and organic acids glucosinolates, alkaloids, polyphenols, tannins) and have been investigated to be efficient in decreasing the corrosion rate of metals in a corrosive area [4, 69]. Inhibitive action of natural compounds is often attributed to some organic constituents of extracts (phytochemicals). These compounds have polar function with N, O, or S atoms, and also conjugated double bonds or aromatic rings in their molecular structure which act as the main center of adsorption and block the active surface sites by adsorption on metal surface, consequently reducing the corrosion rate [35]. Also, the adsorption bond stability relies on the composition of metal and corrodent, inhibitor structure, and concentration [70].

CHAPTER 3

Natural Corrosion Inhibitors: Adsorption Mechanisms

The corrosion inhibiting of natural products can be ascribed to phytochemical substances containing alkaloids, carboxylicacids, ketons, alcohols, ascorbicacids, tannins, nitrogen bases, carbohydrates, proteins, flavonoids, organic pigments, phenolic compounds, aminoacids, and their acid hydrolysis products [55, 71, 72].

Extracts of natural inhibitors like extracts from leaves, barks, seeds, fruits, and roots include mixtures of organic compounds which have nitrogen, sulfur, and oxygen atoms in functional groups (O-H, C=C, C=O, N-H, C=O) as well as multiple bonds, and aromatic rings that act as impressive inhibitors in a corrosive environment [41, 71].

The first step in the inhibition process is the adsorption of banners on the metal surface to prevent occurring corrosion [72]. The various components may react with produced ions on a corroding metal surface like Fe^{2+} on a low carbon steel surface and make organometallic [Fe−Inh] complexes [55]. The adsorption process of an inhibitor occurs with the replacement of one or more water molecules which are preadsorbed on the metal surface:

$$Inh_{sol} + xH_2O_{ads} \longrightarrow Inh_{ads} + xH_2O_{sol}. \qquad (3.1)$$

The subscripts "sol" and "ads" refer to solution and adsorbed species in the order. Then adsorbed inhibitors may combine with Fe^{2+} ions on the metal surface and produce [Fe−Inh] complexes:

$$Fe \longrightarrow Fe^{2+} + 2e \qquad (3.2)$$

$$Fe^{2+} + Inh_{ads} \longrightarrow [Fe - Inh]^{2+}_{ads}. \qquad (3.3)$$

The banning effect of these compounds depends on their resistance and their solubility in the hydrous corrodent. If the complex is unsolved, it means the corrosion reaction is prevented, but if the complex is dissoluble, it will enhance the metal dissolution procedure which it is more obvious at low concentration of the inhibitor. Under this situation, the amount of surface active materials of the inhibitor is not enough to make a protective layer on the surface, so the adsorbed intermediate dissolves fast in a corrosive environment. By increasing the concentration, required organic matter becomes sufficient to establish the complex, reduce solubility, and cause desirable corrosion inhibition [55, 71].

Some popular methods include weight loss, linear sweep voltammetry-LSW (polarization resistance or even more frequently Tafel plot measurements), and electrochemical impedance spectroscopy (EIS) [57].

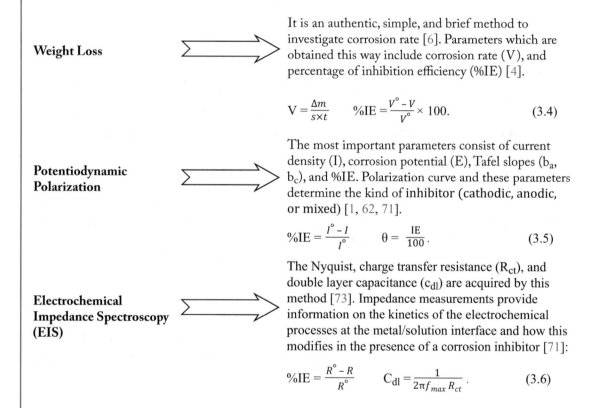

Weight Loss

It is an authentic, simple, and brief method to investigate corrosion rate [6]. Parameters which are obtained this way include corrosion rate (V), and percentage of inhibition efficiency (%IE) [4].

$$V = \frac{\Delta m}{s \times t} \qquad \%IE = \frac{V° - V}{V°} \times 100. \qquad (3.4)$$

Potentiodynamic Polarization

The most important parameters consist of current density (I), corrosion potential (E), Tafel slopes (b_a, b_c), and %IE. Polarization curve and these parameters determine the kind of inhibitor (cathodic, anodic, or mixed) [1, 62, 71].

$$\%IE = \frac{I° - I}{I°} \qquad \theta = \frac{IE}{100}. \qquad (3.5)$$

Electrochemical Impedance Spectroscopy (EIS)

The Nyquist, charge transfer resistance (R_{ct}), and double layer capacitance (c_{dl}) are acquired by this method [73]. Impedance measurements provide information on the kinetics of the electrochemical processes at the metal/solution interface and how this modifies in the presence of a corrosion inhibitor [71]:

$$\%IE = \frac{R° - R}{R°} \qquad C_{dl} = \frac{1}{2\pi f_{max} R_{ct}}. \qquad (3.6)$$

Scheme 3.1: The most widely used methods for measuring corrosion rate.

Inhibition performance depends on the rate of adsorption of its ingredients on a metal surface, but stability of adsorbed molecules (inhibition period) changes with the kind of adsorption, chemical/physical/both, to a great extent. Therefore, it becomes urgent to study metal-inhibitor interaction through adsorption isotherm. Adsorption isotherms which are used to describe adsorption mechanism include Langmuir, Temkin, Frumkin, Florry–Huggins, Freunlich, and Elawady [59]. The main focus of this book is on reporting the influence of natural inhibitors studied by other researchers, not on examing the manner of inhibitor bonding or adsorption isotherm, even though it is mentioned shortly.

CHAPTER 4

Plants as Corrosion Inhibitors in Different Corrosive Environments

4.1 INFLUENCE OF PLANTS IN HCL SOLUTION

As mentioned previously, corrosion inhibitors are used to control corrosion rate and protect corroding surfaces [57]. Behpour et al. [2] evaluated the effect of two oleo-gum resins exudated from ferula assa-foetida (F. assa-foetida) and Dorema ammoniacum (D. ammoniacum) as inhibitors on mild steel in 2.0 M HCl by means of weight loss, potentiodynamic polarization, and EIS. The oleo-gum resin from F. assa-foetida presented higher IE with the oleo-gum resin D. ammoniacum. IE improved with an increase in banner concentration. Banners behaved as mixed inhibitors with major control of an anodic reaction. EIS results showed that addition of inhibitors the double-layer capacitances decrease which refers to the adsorption of these banners on the surface. The adsorption of inhibitors followed the Langmuir isotherm under all investigated temperatures. IE of both banners in 2 M HCl reduced by increasing temperature, and in the presence of both inhibitors activation energy enhanced.

Soltani et al. [6] examined Silybum marianum leaves extract as a natural source inhibitor for 304 stainless steel corrosion in 1.0 M HCl. Weight loss, potentiodynamic polarization, and EIS techniques were applied. S. marianum acted as a mixed inhibitor. At most, banner concentration of $\frac{1g}{l}$ inhibition efficiency was enhanced significantly and became 96%. The adsorption of the inhibitor was automatic and obeyed the Langmuir isotherm. Mixed adsorption (chemical and physical) of an inhibitor was suggested based on thermodynamic parameters. The endothermic nature of the steel dissolution process was detected which means that dissolution of steel was slow. Large and negative values of ΔS^* showed association rather than dissociation.

A comparative study of inhibition by henna and its ingredients (Lawsone, Gallic acid, α-D-Glucose, and tannic acid) on corrosion inhibition of mild steel in 1 M HCl was carried out by Ostovari et al. [7] through weight loss, polarization measurements, and potentiodynamic anodic polarization. It was concluded that all of the tested inhibitors functioned as mixed type. IE increased with a rise of inhibitor concentration to reach a maximum amount of 92.06% at $1.2\,gl^{-1}$ henna extract. IE grew in the order: lawsone > gallic acid > α-D-Glucose (dextrose) > tannic acid. Chemisorptions mechanism was suggested for the adsorption of banner molecules on the metal surface. Lawsone was found to be the main element which was responsible for the corrosion inhibition of mild steel.

Jasminum nudiflorun lindl leaves extract (JNLLE) acted as a good inhibitor for corrosion of Al in 1 M HCl solution. Deng and Li [35] investigated inhibition efficiency of JNLLE by means of weight loss, polarization curves, EIS, and SEM techniques. JNLLE acted as a cathodic banner. IE increased by banner concentration and reduced by temperature. Higher than 90% IE was achieved at 20°C and 2 h immersion time. IE enhanced by extending immersion time up to 2 h and then little by little decreased. IE also flourished by acid concentration from 0.5–1 M but then diminished. Adsorption of JNLLE on a metal surface followed the Langmuir isotherm. The process of adsorption was physically and exothermic along with a decrease in entropy. The presence of JNLLE increased R_t values while decreased C_{dl} values.

The corrosion inhibition of 304 stainless steel in 2.0 M HCl by Ferula gumosa (galbanum) extract as a natural inhibitor was tested by Behpour et al. [74]. Weight loss, potentiodynamic polarization, and EIS methods were applied. IE increased with increasing galbanum extract (GE) concentration and rising temperature. The tafel polarization technique showed GE functioned as a mixed inhibitor with significant control of anodic reaction. The main substances include β-pinene and α-pinene which carried out corrosion inhibition. Other ingredients like δ-3-carene and limonene, sesquiter-phenols, esters, coumarins, carboxylicacids, diazines, and furanoids may amplify the power of the film made over the steel surface, so these phytoconstituents suggested protection through adsorption.

The adsorption of banner on the metal obeyed Temkin adsorption isotherm. Soltani et al. [75] evaluated the effect of Salvia officinalis (S. Officinalis) extract as a natural inhibitor for 304 stainless steel in 1.0 M HCl solution by methods of weight loss, potentiodynamic polarization, and EIS. S. officinalis acted as a good and mixed inhibitor. The IE increased with inhibitor concentration. The adsorption process was automatic and obeyed the Langmuir isotherm at all considered temperature. The values of ΔG°_{ads} were representative of the physisorption. Activation energy diminished with the addition of banner.

The inhibition action of carmine and fast green dyes as inhibitors on corrosion of mild steel was studied in 0.5 M HCl by Enkatesha et al. [76]. It used mass loss, polarization, and EIS methods. Inhibitors acted as good and mixed type with significant cathodic effect. IE increased by a rise in banners concentration. Optimum concentration of banners was 1×10^{-3} M solution at 300 K for which the IE was maximum. The IE of fast green (IE: 98%) was more than carmine (IE: 92%). The adsorption of inhibitors followed the Temkin isotherm.

The influence of some natural product extracts as eco-friendly corrosion inhibitors including Hybiseus Syriacus Linn and the flavonoid component Rutin for mild steel in 1 M HCl was studied by Nagarajan et al. [77]. It applied weight loss, gasometric, and polarization methods in the presence and absence of quaternaryammonium salt. The extract of plant and rutin decreased the corrosion rate. It was observed that plant extract was a more effective inhibitor thanks to the presence of active inhibiting compounds. IE increased with a rise in banner concentration.

Khalaf et al. [78] investigated the effects of aqueous extract of Gummara of the Phoenix Dactylifera (AEGPD) and aqueous extract of Lactuca (AEL) on the corrosion inhibition of mild steel in a 1 M HCl solution by weight loss measurements at a temperature range of 30–60°C. Results showed inhibition percentage enhanced with an increase of inhibitor concentration and decreased by a rise in temperature. AEGPD and AEL were revealed to be highly efficient inhibitors. IE obtained 93.82% and 95.81% at the concentration of $2 \frac{g}{l}$ at 30°C for AEGPD and AEL, respectively. The adsorption of inhibitors obeyed the Langmuir isotherm. Spontaneous adsorption and the physisorption process was confirmed for both banners.

Hussin and Kassim [79] investigated the inhibitory action of (+)-catechin hydrate on the corrosion of mild steel in 1 M HCl solution by tools of weight loss, potentiodynamic polarization, and electrochemical impedance spectroscopy. Results showed (+)-catechin hydrates functioned as a good and a mixed type inhibitor with significant inhibition at the anodic site. Corrosion of mild steel in the absence and presence of inhibitors controlled by the charge transfer process. The IE of mentioned banner decreased with temperature. The inhibitive action of banners followed the Langmuir isotherm. The adsorption process of inhibitors on the metal surface was automatic, and physical.

It was discovered that natural polymer Xanthan Gum (XG) acted as an eco-friendly corrosion inhibitor for mild steel in 1 M HCl at 30°C, 40°C, 50°C, and 60°C. Mobin and Rizvi [80] investigated the inhibition effect of XG on mentioned metal. Researchers applied gravimetric analysis, potentiodynamic polarization, EIS, chemical calculations, SEM, and UV-visible spectrophotometry to do an experiment. XG notably decreased the corrosion rate of mild steel and acted as a good and mixed inhibitor. IE increased by the addition of XG and diminished by a rise in temperature. IE increased synergistically with the addition of surfactants including sodium dodecyl sulfate (SDS), cetyl pyridinium chloride (CPC), and Triton X-100. SDS was more effective than others. IE of additives was in order: XG + SDS > XG + Triton X100 > XG + CPC > XG. The formation of an insoluble protective layer on the metal surface in the presence of XG and XG-Surfactants admixtures were confirmed by EIS method. The corrosion process was hindered by adsorption of XG on the metal surface obeying the Langmuir isotherm.

The performance of creatinine as a corrosion inhibitor for mild steel in 1 M HCl solution was studied by Al-Amiery et al. [81] who applied the weight loss method. The inhibitory action occurred by adsorption of the creatinine molecules on the metal surface. As the results revealed, the mentioned banner acted as a very good inhibitor. By the growth of inhibitor concentration, IE increased and reached 95.1% at the highest concentration. IE also reduced by a rise in temperature which means the adsorption of banner follows Langmuir.

The inhibitory action of Ilex Paraguariensis (green Yerba mate) extracts on the corrosion of carbon steel in 1 M HCl solution was examined by D'Elia et al. [82]. Open circuit potential, potentiodynamic polarization curves, weight loss, EIS, and SEM methods were done. Green mate extract acted as a mixed inhibitor on the C-steel surface. By rising extract concentration and immersion time, IE enhanced. It remained constant with changing temperature. The examined inhibitor adsorbed on the C-steel surface chemically, because of decreasing apparent activation energy for dissolution of C-steel surface in the presence of extract. The adsorption of the inhibitor followed the Langmuir isotherm. The formation of a soft surface on C-steel in the presence of Yerba mate extract compositions was proved by SEM analysis which was likely due to the formation of a potent chemisorptive bond between yerba mate extract compounds and C-steel surface. In addition, the inhibitory action of natural products was not only attributed to the phenolic matters.

Chidiebere et al. [83] examined the IE of Funtumia Elastica (FE) extracts on corrosion of Q235 mild steel in 1 M HCl solution by some methods including potentiodynamic polarization and EIS. The results showed FE inhibited corrosion of mild steel effectively via adsorption of organic matter on the metal/solution interface, and functioned as a mixed inhibitor, influencing both cathodic and anodic reactions. The adsorption of the inhibitor was validated by the Langmuir isotherm. Chemical adsorption occurred by FE on the corroding metal surface.

Almayouf et al. [84] studied anticorrosive properties of various extracts (methanolic, aqueous methanolic, and water extracts) of Anthemis Pseudocotula for mild steel in 1 M HCl. Methanolic extract of A. Pseudocotula revealed the highest anticorrosive effect among the different examined extracts. Anticorrosive assay-guided isolation of this methanolic extract caused in the isolation of a highly powerful anticorrosive compound, ABP (Luteolin-7-o-β-D-glucoside) which its inhibition effect on corrosion of mild steel was explored utilizing gravimetric, tafel plots, linear polarization, EIS, SEM, and EDS. According to achieved results, ABP inhibited the corrosion of mild steel. IE significantly enhanced by rising in concentration of ABP. ABP acted as a mixed type predominantly cathodic effect. The adsorption of inhibitor on mild steel obeyed the Langmuir isotherm, and the process of adsorption was spontaneous, exothermic, and physical.

Prabhu et al. [85] evaluated the IE of Azadirachta indica (AI-Neem), a natural product on the corrosion of zinc in 2.0 M HCl solution by means of mass loss, electrochemical polarization, and impedance methods. AI-Neem was detected to be a fairly good inhibitor for zinc in HCl solution. IE increased with concentration of banner and reached to its maximum at 1,000 ppm. AI acted as a mixed inhibitor with mainly cathodic effect. Temkin isotherm was suggested for the adsorption of AI-Neem on the zinc surface.

IE of launaea nudicaulis methanolic (MLN) and aqueous acidic (ALN) extracts as a source of new and efficient green inhibitor on corrosion of mild steel in 1 M HCl, was investigated by Alkhathlan et al. [67] applying weight loss, tafel plots linear polarization, EIS, as well as SEM, and EDS. Both extracts (MLN and ALN) blocked notable corrosion of mild steel in 1 M HCL. IE increased by a rise in concentration of extracts, and reached 92.5% for MLN and 87.2% for ALN. MLN and ALN acted as mixed-type inhibitors with the main anodic effect. Adsorption behavior of both green inhibitors obeyed the Langmuir isotherm. The adsorption process of both banners was spontaneous, exothermic, and physical. Results represent that L. Nudicaulis can present effective inhibition for mild steel in an HCL environment.

Myrrh gum and magnetite/Myrrh nanocomposites were found to have inhibiting properties for corrosion of steel in 1 M HCl by Atta et al. [86] using potentiodynamic polarization and EIS methods. Myrrh and magnetite/Myrrh nano particles behaved as mixed-type inhibitors because of inhibiting both cathodic and anodic reactions. Corrosion resistance enhanced by increasing inhibitor concentration. The maximum value of IE was achieved at 91%.

Lei et al. [87] evaluated the aqueous Michelia alba leaves extract (MALE) as an inhibitor to corrosion of various steel materials (industrial pure iron, stainless steel, and carbon steel) in 1 M HCl by means of potentiodynamic polarization, SEM, and FT-IR. Results indicated that MALE played as a highly impressive mixed inhibitor for all mentioned steels and rising temperature benefited its corrosion inhibition. The adsorption of MALE on three steel surfaces followed the Langmuir isotherm. The component liriodenine took the most important role in the inhibition performance of MALE.

The behavior of C-steel in 1 M HCl solution in the presence of chenopodium Ambrorsioides extract (CAE) was studied by Salghi et al. [88] who applied weight loss, potentiodynamic polarization curves, and EIS methods. According to acquired information, it was found that CAE performed as an excellent and cathodic inhibitor. IE increased with inhibitor concentration and temperature. The adsorption of CAE on the metal surface followed the Langmuir isotherm.

Nasibi et al. [89] used chamomile (Matricaria recutita) extract (CE) as a corrosion inhibitor for mild steel in 1 M HCl solutions by electrochemical (polarization and EIS) methods and surface analysis (optical microscopy/AFM/SEM) studies. Electrochemical investigation revealed CE functioned as a mixed kind influencing mainly anodic action. Maximum inhibition efficiency was obtained at 93.2% applying 7.2 gl^{-1} inhibitor at 22°C. IE reduced by increasing temperature and pH. CE adsorbed on the metal surface physically. The mechanism of adsorption was in the harmony with the Langmuir isotherm.

Inhibition properties of natural Junipers extract (JE) on the corrosion of C38 steel in 1 M HCl solution were determined by Salghi et al. [90]. Mass loss, potentiodynamic polarization curves, and EIS were used. JE was proved to have excellent IE and were classified as a mixed-type inhibitor. IE increased with concentration. The adsorption process was spontaneous and exothermic. It also obeyed the Langmuir isotherm.

Mangrove Tannin (MT) was used as corrosion inhibitor on corrosion of copper in 0.5 M HCl solution by Rahim et al. [91], weight loss method, potentiodynamic polarization, EIS, SEM, accompanied with EDX, atomic absorption spectroscopy (AAS), and ion chromatography (IC) were applied to investigate inhibition features of MT. MT played primarily as cathodic inhibitors, so influenced the cathodic process, reduced corrosion current density, and moved the corrosion potential toward more negative amounts. By addition of 3 gl^{-1} MT, IE reached 82%. The adsorption of MT on a copper surface fitted well with the Langmuir isotherm. According to results obtained from AAS and IC, the addition of MT reduced diffusion of copper (II) ions and enhanced Cl^{-1} ion concentrations in the electrolyte solution after corrosion reaction.

A comparative study of corrosion inhibition of three green compositions named Argonia spinosa press cake, Kernels, and Hulls extracts on mild steel in a 1 M HCl medium was done by Afia et al. [92]. Each of the three compounds were excellent inhibitors for steel in an acidic medium at 298 K and functioned as mixed-type inhibitors by influencing both anodic and cathodic reactions. IE increased by a rise of the concentration of inhibitors. The adsorption of three compounds on a steel surface followed the Langmuir isotherm. The information acquired with various techniques including EIS and potentiodynamic polarization were in good agreement.

The aqueous extract of thyme leaves was examined thanks to its ability as a corrosion barrier for mild steel in 2 M HCl by Ibrahim et al. [93]. Weight loss measurements, linear polarization resistance (LPR), EIS, and cyclic sweep (CP), as well as FT-IR, were used. Based on conclusions, thyme leaf extract behaved as a mixe-type with mainly anodic efficiency. IE increased with the rise of banner concentration. Corrosion rate increased with higher temperature which means physical adsorption of an inhibitor on a metal surface and mechanism of adsorption obeyed the Langmuir isotherm.

Salghi et al. [94] investigated the corrosion inhibition of Argan Kernels extract (AKE) and Cosmetic Argan Oil (CAO) on mild steel in 1 M HCl by means of gravimetric and electrochemical techniques including potentiodynamic polarization and EIS. Results showed AKE and CAO acted as mixed-type inhibitors without modifying the mechanism of hydrogen evolution. IE increased with banner concentration and diminished with rise of temperature. At $1\frac{g}{l}$ of KAE and $6\frac{g}{l}$ of CAO, IE reached 97% and 91%, respectively. The adsorption of mentioned banners was according to the Langmuir isotherm.

The effect of the extract of Azadirachta excelsa leaves as a corrosion barrier on mild steel in 1 M HCl was determined using weight loss method by Mahat et al. [95]. It was proved that corrosion rate decreased and IE enhanced as the inhibitor was added. With the highest concentration of mentioned banner, the surface was soft and less corroded. All of the obtained results revealed Azadirachta excelsa has an identified green inhibitor.

The study of the effect of black pepper (BP) extract and its piperine isolated from BP on corrosion of C38 steel in 1 M HCl solution was done by Hammouti et al. [96] utilizing the weight loss method. BP and piperine had impressive inhibition and acted as an effective inhibitor. The maximum value of IE was achieved to 95.8% at $2\frac{g}{l}$ of BP. IE for piperine was attained at 99% at 10^{-3}M. Piperine was adsorbed on the metal surface based on the Langmuir isotherm with likely chemisorptions phenomenon.

Aqueous extract of coffee grounds can act as an impressive inhibitor for C-steel in a 1 M HCl solution. D'Elia et al. [97] applied weight loss method, electrochemical measurements, including potentiodynamic polarization, and EIS. According to achieved conclusions, examined extracts acted as a mixed type with mainly a cathodic effect. IE enhanced by rising temperature. The inhibitory action of an inhibitor was imputed to a potent chemisorptive bond. The Langmuir isotherm was fitted for the adsorption of a banner.

Rajalakshmi et al. [98] examined the corrosion mitigation effect of cyamopsis tetrago-naloba seed extract on mild steel in a 1 M HCl solution by means of weight loss method and potentiodynamic polarization technique. The tested inhibitor performed as a mixed mode of inhibition. 92% IE was provided at 0.7% concentration. The adsorption of banner obeyed the Langmuir and Temkin isotherms.

Ibrahim and Habbab [99] tested inhibition action of eggplant peel extract on corrosion of mild steel in 2 M HCl by employing weight loss and electrochemical techniques such as LPR, EIS, and CS. Researchers concluded eggplant peel extract acted as a mixed-type inhibitor. IE increased by the concentration of inhibitor and reached 84%. The Langmuir isotherm was described for the adsorption of the banner.

Traditional weight loss and electrochemical techniques including LPR, CS, and AC EIS were applied to examine corrosion inhibition of mild steel using fig leaf extract in 2 M HCl by Ibrahim and Abou Zour [100]. IE was obtained as high as 87–90%. Adsorption of the inhibitor obeyed the Langmuir isotherm. This process was spontaneous and physical.

Goel et al. [101] isolated three pyridine base alkaloids named ricinine (C1), N-demethylricinine (C2), and 4-methoxypyridine-3-carboxylic acid (C3) from methanolic extract of Riccinus Communis leaves and studied the effects of inhibition on corrosion of mild steel in 0.5 M HCl solution using weight loss and electrochemical methods (Galvanostatic polarization and EIS). Based on obtained conclusions, C1 was an excellent barrier for mild steel in an acidic medium. C1, C2, and C3 were found to be mixed inhibitors. IE increased by concentration of barriers and followed the order: C1 > C2 > C3. But it decreased with rising temperature. The adsorption of three mentioned inhibitors obeyed the Langmuir isotherm. It was also sponta-neous and exothermic.

Piperanine was isolated from BP and tested as eco-friendly corrosion inhibitors for C38 steel in 1 M HCl solution by Hammouti et al. [102] applying the weight loss method at different temperatures. Piperanine diminished corrosion rate of C38 steel. IE increased by rising inhibitor concentration and the maximum value was achieved at 97.5% in $10^{-3} \frac{mol}{l}$ but it decreased with an increase of temperature. Piperanine adsorbed on the metal surface by both physisorption and chemisorption following the Langmuir isotherm.

Corrosion inhibitive of mild steel in 2 M HCl was examined by Ibrahim et al. [103] using potato peel extract (PPE). According to the obtained results from weight loss and electrochem-ical techniques such as LPR, CS, and EIS, PPE provided excellent inhibitive properties and behaved as a mixed-type barrier. IE was achieved as high as 85–90% with various concentra-tions of PPE, however, when concentration of PPE reached optimum levels barrier molecules remained in the bulk solution and no notable adsorption occurred. The Langmuir isotherm was fitted for the adsorption of PPE on the metal surface.

Corrosion inhibition of carbon steel in 1 M HCl solution with extract of brown alga Bi-furcaria bifurcata (Bb) was studied by Abboud et al. [104] using weight loss, potentiodynamic, and polarization resistance measurements. It was found that Bb played as a mixed inhibitor mainly as an anodic banner at higher concentration. In addition to the conservation influence of Bb extract by means of adsorbing on the metal surface, FT-IR and UV-visible studies revealed inhibition may also be thanks to the formation of an insoluble film through the process of complexation of the Bb extract on the metal surface.

Testing natural extract of fenugreek on the corrosion of steel in 1 M HCl was carried out by Bouyanzer et al. [105] applying weight loss, polarization, and EIS measurements. According to the obtained results, fenugreek acted as a cathodic inhibitor. IE flourished with inhibitor concentration and reached 94% at 10 gl^{-1} of extract. It was independent of temperature.

Ginseng root, a new impressive corrosion inhibitor, for Al alloy of type AA1060 in 1 M HCl solution was used by Obot and Obi-Egbedi [106]. Weight loss measurements were carried out. It was found that the mentioned inhibitor worked as impressive and efficient for Al corrosion in an HCl solution. Corrosion rate enhanced in temperature both in free form and containing an inhibitor as well as reduced in the presence of the extract. IE of Ginseng extract increased with a rise of inhibitor concentration and reduced with a rise of temperature reaching its maximum (93.1) at 50% $\frac{V}{V}$ concentration of Ginseng. The addition of iodide ions caused an enhancement of IE. The adsorption of the inhibitor was found to be spontaneous, physical, and obeyed the Freundlich isotherm.

Corrosion inhibition by Justicia gendaruss a plant extract (JGPE) on mild steel in 1 M HCl was examined by Gunasekaran et al. [107]. Weight loss, potentiodynamic polarization, and EIS methods were used. JGPE acted as a mixed-type inhibitor without modifying the mechanism of hydrogen evolution. IE increased with JGPE concentration up to 93% at 150 ppm at 25°C. In the range of 25–75°C, JGPE was impressive in decreasing corrosion of mild steel. AFM and XPS studies exhibited the formation of banner layer including JGPE molecules, chloride, and iron oxides. Physical adsorption and the Langmuir isotherm were suggested for the adsorption of JGPE on mild steel.

Umoren and Ebenso [108] examined the anti-corrosive effect of Raphia hookeri exudate gum and halide ions on corrosion inhibition of Al in 0.1 M HCl solution by means of gravimetric and gasometric techniques. According to finding results, IE increased with inhibitor concentration and decreased with temperature. Inhibition abilities of the exudates enhanced in the presence of halide ions thanks to the synergistic effects in the order: KI > KBr > KCL. Inhibition action of a mentioned barrier was related to the adsorption of phytochemical ingredients. The adsorption of the inhibitor was proposed to be physical, and Freundlich, Langmuir, as well as Temkin isotherm were described for that.

Khillah (Ammi visnaga) seed extract was tested as an inhibitor for corrosion of SX 316 steel in 2 M HCl solution by El-Etre [109] applying weight loss and potentiostatic polarization methods. It was concluded that Khillah includes furanochromones (Khellin and visnagin) which were responsible for inhibition action and performed by adsorption on metal steel. The adsorption of inhibitor obeyed the Langmuir isotherm and was spontaneous. IE increased with a rise of Khillah concentration and diminished with a rise of temperature.

Mahmoud [110] determined the IE of water extracts of some plants including outer brown skin of onion (A), onion bulb (B), the cloves of garlic bulb (C), orange peels (D), and henna leaves (E) on the corrosion of Muntz (63% cu, ≈37% Zn) alloy in 1 M HCl by the use of some methods such as weight loss, galvanostatic polarization, linear polarization, and atomic adsorption spectroscopy. Based on the results, explored extract had a high protection performance on the corrosion of Muntz. IE depended on the kind and concentration of the extracts reducing in the order: C > D > E > B > A. All mentioned extracts functioned as a mixed inhibitor affecting both cathodic and anodic reactions. The adsorption of these extracts obeyed Frumkin's isotherm. The atomic adsorption spectroscopy exhibited the existence of mentioned extracts significantly hindered the preferential dissolution of zinc from the alloy in the belligerent solution.

Table 4.1 summarizes the influence of plants in HCl media.

4.2   EFFECT OF PLANTS IN H$_2$SO$_4$ MEDIA

The effect of Musa Paradisiaca extract on corrosion inhibition of mild steel in aqueous 0.5 M H$_2$SO$_4$ was tested by Mayanglambam et al. [111]. Weight loss method, potentiodynamic polarization, and EIS techniques were applied to investigate the inhibition effect of Musa Paradisiaca. The mentioned inhibitor functioned as a mixed kind inhibitor. IE increased with the concentration of barrier and reduced by rising temperature. The adsorption of molecules of banner on the metal surface obeyed the Langmuir isotherm. Musa Paradisiaca made a physical obstacle because the values of activation energies of the corrosion process increased in the presence of banner which means a reduction in corrosion rate of mild steel in 0.5 M sulfuric acid. The negative values of $\Delta G°_{ads}$ and $\Delta H°_{ads}$ presented spontaneous and exothermic adsorption on the metal surface. It was also found chemical and physical adsorption for the process.

IE of solanum nigrum as a natural source on corrosion of mild steel in 1 M H$_2$SO$_4$ was studied by Sethuraman and Raja [112]. Potentiodynamic polarization, AC impedance, and SEM techniques were utilized. It was revealed the mixed kind of inhibition by influencing both cathodic and anodic reactions. The anticorrosive effect of the mentioned barrier was ascribed to the presence of solasodine. IE increased by rising concentration of plant extract as well as temperature. Chemisorption mechanism following Temkin isotherm was suggested for the adsorption of barrier on metal surface.

Adsorption and inhibition properties of ethanol extract of Gongronema Latifolium as corrosion inhibitor for mild steel in 2 M H$_2$SO$_4$ were investigated by Eddy and Ebenso [113]. IE was appraised utilizing thermodynamic and gasometry methods. It differed with concentration of barrier, immersion time, and temperature. Inhibition properties of ethanol extract of GL was thanks to the presence of amino acid, tannin, carotenoid, flavanoid, steroid, and anthrocyanin in the extract. Physical mechanism and the Langmuir isotherm were suggested for the adsorption of barrier on the metal surface. Inhibition characterization of natural extract of Artemisia on the corrosion of steel in 0.5 M H$_2$SO$_4$ in the temperature range 298–353 K was investigated by means of electrochemical polarization and linear polarization measurements.

Table 4.1: Summary of influence of plants in HCl media (*Continues*)

NO	Metal	Inhibitor	Additives	Methods	Findings	Ref
1	Mild steel	Two oleo-gum resins exudate from ferula assa-foetida (F. assa-foetida) and Dorema ammoniacum (D. ammoniacum)		Weight loss, potentiodynamic polarization, EIS	Mixed-mode inhibitors with major anodic effect, higher IE for oleo-gum resin from f. assa-foetida than oleo-gum resin D. ammoniacum, reduction of IE with rise of temperature, Langmuir isotherm	[2]
2	304 stainless steel	Silybum marianum leaves		Weight loss, potentiodynamic polarization, EIS	Mixed-type inhibitor, Max of IE: 96%, Langmuir isotherm, mixed adsorption (physical and chemical), endothermic solution	[6]
3	Mild steel	Henna and its ingredients (Lawsone, Gallic acid, α-D-Glucose and tannic acid)		Weight loss, polarization, potentiodynamic anodic polarization	Mixed-type banner, the maximum amount of IE: 92.06% at 1.2 gl^{-1} henna extract, IE of inhibitors: lawsone> gallic acid > α-D-Glucose (dextrose) > tannic acid, Chemisorptions adsorption mechanism	[7]
4	Aluminum (Al)	Jasminum nudiflorun lindl. Leaves extract (JNLLE)		Weight loss, polarization curves, EIS, SEM	Cathodic banner, reduction of IE by temperature, Higher than 90% inhibition efficiency at 20°C and 2 h immersion time, Langmuir isotherm, physical and exothermic adsorption process along with a decrease in entropy	[35]
5	304 stain less steel	Ferula gumosa (galbanum)		Weight loss, potentiodynamic polarization, EIS	Mixed mode inhibitor, enhancement of IE by rise of inhibitor concentration and temperature, β-pinen and α-pinene the main active components, Temkin adsorption isotherm	[74]

Table 4.1: (*Continued*) Summary of influence of plants in HCl media (*Continues*)

6	304 stain less steel	Salvia officinalis (S. officinalis)		Weight loss, potentiodynamic polarization, EIS	Mixed-type inhibitor, physical adsorption, Langmuir isotherm,	[75]
7	Mild steel	Carmine and fast green dyes		Mass loss, polarization, EIS	Mixed-type inhibitors with significant cathodic effect, higher IE for fast green than carmine, Temkin isotherm	[76]
8	Mild steel	Hybiseus Syriacus Linn and the flavonoid component Rutin	Quaterna-ryammonium salt	Weight loss, gasometric, polarization	Reduction of corrosion rate because of functional groups in aromatic and hetrocyclic rings accompanied with electron rich elements like nitrogen, sulphur, and oxygen	[77]
9	Mild steel	Gummara of the Phoenix Dactylifera (AEGPD), Lactuca (AEL)		Weight loss at temperature range (30-60°C)	Reduction of IE with rise of temperature, physical and spontaneous adsorption, Langmuir isotherm	[78]
10	Mild steel	(+)-Catechin hydrate		Weight loss, potentiodynamic polarization, electrochemical impedance spectroscopy, SEM	Mixed mode inhibitor, reduction of IE with rise of temperature, physical adsorption, Langmuir isotherm	[79]
11	Mild steel	Natural polymer Xanthan Gum (XG)	Sodium dodecyl sulfate (SDS), cetyl pyridinium chloride (CPC), and Triton X-100	Gravimetric analysis, potentio-dynamic polariza-tion, EIS, chemical calculations, scanning electron microscopy (SEM), and UV-visible spectrophotome-try, at 30°C, 40°C, 50°C, 60°C	Mixed inhibitor, reduc-tion of IE by rising tem-perature, enhancement of IE with rise of con-centration of inhibitor and addition of surfac-tants in the order: XG + SDS > XG + Triton X100 > XG + CPC > XG, physical adsorption, Langmuir isotherm	[80]

Table 4.1: (*Continued*) Summary of influence of plants in HCl media (*Continues*)

12	Mild steel	Creatinine		Weight loss, SEM	Increase of IE with rise of barrier concentration and reduction of that with rise of temperature, maximum of IE: 95.1%, Langmuir isotherm	[81]
13	Carbon steel	Ilex Paraguariensis (green Yerba mate)		Open circuit potential, potentiodynamic polarization curves, weight loss, EIS, SEM	Mixed inhibitor, enhancement of IE by rising extract concentration and immersion time also remained constant with changing temperature, chemical adsorption, Langmuir isotherm	[82]
14	Q235 mild steel	Funtumia Elastica (FE)		Potentiodynamic polarization, EIS	Mixed inhibitor, chemical adsorption, Langmuir isotherm	[83]
15	Mild steel	Extracts (methanolic, aqueous methanolic, and water extracts) of Anthemis Pseudocotula		Gravimetric, tafel plots, linear polarization, EIS, SEM, EDS	The highest anticorrosive effect of methanolic extract of A. Pseudocotula among the different examined extracts, ABP (Luteolin-7-o-β-D-glucoside), highly powerful anticorrosive compound, mixed-type predominantly cathodic effect (ABP), spontaneous, exothermic and physical adsorption, Langmuir isotherm	[84]
16	Zinc	Azadirachta indica (AI-Neem)		Mass loss, electrochemical polarization, impedance	Mixed inhibitor with mainly cathodic effect, Temkin isotherm	[85]
17	Mild steel	Launaea nudicaulis methanolic (MLN) and aqueous acidic (ALN)		Weight loss, tafel plots linear polarization, EIS, as well as SEM, EDS	Mixed-type inhibitors with main anodic effect, maximum of IE 92.5% for MLN and 87.2% for ALN, exothermic, spontaneous, and physical adsorption, Langmuir isotherm	[67]

Table 4.1: (*Continued*) Summary of influence of plants in HCl media (*Continues*)

18	Steel	Myrrh gum and magnetite/Myrrh nanocomposites		Potentiodynamic polarization EIS	Mixed-type inhibitors, the maximum value of IE: 91%	[86]
19	Various steel materials (industrial pure iron, stainless steel, and carbon steel)	Michelia alba leaves (MAL)		Potentiodynamic polarization, SEM, FT-IR	Mixed inhibitor, liriodenine, the most important compound in inhibition action, Langmuir isotherm	[87]
20	C-steel	Chenopodium Ambrorsioides (CA)		Weight loss, potentiodynamic polarization curves, EIS	Cathodic inhibitor, increase of IE with inhibitor concentration and temperature, Langmuir isotherm	[88]
21	Mild steel	Chamomile (Matricaria recutita)		Polarization EIS, optical microscopy/AFM/SEM	Mixed kind influencing mainly anodic action, Maximum inhibition efficiency 93.2% applying 7.2 gl⁻¹ inhibitor at 22°C, reduction of IE by increasing temperature and pH, physical adsorption, Langmuir isotherm	[89]
22	C38 steel	Junipers extract (JE)		Mass loss, potentiodynamic polarization curves, EIS	Mixed type inhibitor, spontaneous and exothermic adsorption process, Langmuir isotherm	[90]
23	Copper	Mangrove Tannin (MT)		Weight loss, potentiodynamic polarization, EIS, SEM, accompanied with EDX, atomic absorption spectroscopy (AAS), ion chromatography (IC)	Cathodic inhibitor, Langmuir isotherm	[91]

Table 4.1: (*Continued*) Summary of influence of plants in HCl media (*Continues*)

24	Mild steel	Argonia spinosa press cake, Kernels, and Hulls		EIS, potentiodynamic polarization	Mixed-type inhibitors influencing both anodic and cathodic reactions, Langmuir isotherm	[92]
25	Mild steel	Thyme leaves		Weight loss, linear polarization resistance (LPR), EIS, and cyclic sweep (CP), also FT-IR	Mixed-type with mainly anodic efficiency, physical adsorption, Langmuir isotherm	[93]
26	Mild steel	Argan Kernels extract (AKE) and Cosmetic Argan Oil (CAO)		Gravimetric and electrochemical techniques including potentiodynamic polarization and EIS	Mixed-type inhibitors without modifying the mechanism of hydrogen evolution, reduction of IE with rise of temperature, Langmuir isotherm	[94]
27	Mild steel	Azadirachta excelsa leaves		Weight loss	Reduction of corrosion rate and enhancement of IE with addition of inhibitor	[95]
28	C38 steel	Black pepper (BP), Piperine isolated from BP		Weight loss	The maximum value of IE: 95.8 at $2\frac{g}{l}$ of BP and 99% at 10^{-3}M Piperine, Langmuir isotherm with likely chemisorptions phenomenon for the adsorption of piperine	[96]
29	C-steel	Coffee grounds		Weight loss method, electrochemical measurements including potentiodynamic polarization, EIS	Mixed-type inhibitor with mainly cathodic effect, enhancement of IE by rising temperature, Langmuir isotherm	[97]
30	Mild steel	Cyamopsis tetragonaloba seed		Weight loss, potentiodynamic polarization	Mixed mode, 92% inhibition efficiency at 0.7% concentration of inhibitor, Langmuir and Temkin isotherms	[98]
31	Mild steel	Egg plant peel		Weight loss, electrochemical techniques such as LPR, EIS,CS	Mixed type, the maximum value of IE: 84%, Langmuir isotherm	[99]

Table 4.1: (*Continued*) Summary of influence of plants in HCl media (*Continues*)

32	Mild steel	Fig leaves		Weight loss, electrochemical techniques including LPR, CS, AC EIS	As high as 87-90% inhibition efficiency, Langmuir isotherm, spontaneous and physical adsorption	[100]
33	Mild steel	Pyridine base alkaloids named ricinine (C1), N-demethylricinine (C2), and 4-methoxypyridine-3-carboxylic acid (C3) from methanolic extract of Riccinus Communis leaves		Weight loss, electro-chemical methods (galvanostatic polarization and EIS)	Mixed inhibitors, inhibition efficiency of banners: C1 > C2 > C3, reduction of IE with rising temperature, Langmuir isotherm, spontaneous and exothermic adsorption	[101]
34	C38 steel	Piperanine isolated from BP		Weight loss method at different temperatures	The maximum value of IE: 97.5% at $10^{-3}\frac{mol}{l}$, physisorption and chemisorption following Langmuir isotherm	[102]
35	Mild steel	Potato peel extract (PPE)		Weight loss and electrochemical techniques such as LPR, CS, EIS	Mixed-type barrier, as high as 85-90% IE, Langmuir isotherm	[103]
36	Carbon steel	Brown alga Bifurcaria bifurcata (Bb)		Weight loss, potentiodynamic, and polarization resistance, FT-IR and UV-visible	Mixed inhibitor mainly as an anodic banner at higher concentration	[104]
37	Steel	Fenugreek		Weight loss, polarization, EIS	Cathodic inhibitor, maximum value of IE: 94% at 10 gl^{-1} of extract, IE was independent of temperature	[105]

Table 4.1: (*Continued*) Summary of influence of plants in HCl media (*Continues*)

38	Aluminium alloy of type AA1060	Ginseng root	Iodide ions	Weight loss	Increase of inhibition efficiency of Ginseng extract with rise of inhibitor concentration and reduction with rise of temperature, the maximum value of IE: 93.1 at 50% $\frac{v}{v}$ concentration of Ginseng, an enhancement of inhibition efficiency by addition of iodide ions, spontaneous and physical adsorption, Freundlich isothem	[106]
39	Mild steel	Justicia gendaruss a plant extract (JGPE)		Weight loss, potentiodynamic polarization, EIS, AFM, XPS	Mixed-type inhibitor without modifying the mechanism of hydrogen evolution, increase of IE with JGPE concentration up to 93% at 150 ppm at 25°C, physical adsorption and Langmuir isotherm	[107]
40	Aluminium	Raphia hookeri exudate gum	Halid ions	Gravimetric and gasometric	Decrease of IE with temperature, increase of inhibition abilities of the exudates in the presence of halide ions thanks to the synergistic effects in the order: KI>KBr>KCL, physical adsorption, Freundlich, Langmuir, and Temkin isotherm	[108]
41	SX 316 steel	Khillah (Ammi visnaga) seeds		Weight loss, potentiostatic polarization	Furanochromones (Khellin and visnagin), responsible for inhibition action, Langmuir isotherm, spontaneous adsorption, increase of IE with rise of Khillah concentration and reduction of that with rise of temperature	[109]

Table 4.1: (*Continued*) Summary of influence of plants in HCl media

42	Muntz (63% cu, ≈37% Zn) alloy	Water extracts of some plants including outer brown skin of onion (A), onion bulb (B), the cloves of garlic bulb (C), orange peels (D), henna leaves (E)		Weight loss, galvanostatic polarization, linear polarization, atomic adsorption spectroscopy	Mixed inhibitor affecting both cathodic and anodic reactions, IE depending on the kind and concentration of the extracts: C> D> E> B> A, Frumkin's isotherm	[110]

Hammouti and Bouklah [114] concluded that Artemisia played as a mixed inhibitor without changing the mechanism of the hydrogen evolution reaction. IE enhanced with Artemisia concentration and temperature to attain 95% as well as 99% at 10 gl^{-1} at 298 and 353 K, respectively. Langmuir isotherm was described for the adsorption of Artemisia.

Protective properties of Chenopodium Ambrosioides extract (CAE) on corrosion of carbon steel in 0.5 M H$_2$SO$_4$ solution were tested by Salghi et al. [115] with the help of weight loss, potentiodynamic polarization, and EIS techniques. Based on the results, CAE acted as mainly as a cathodic inhibitor. IE increased with a rise of concentration of CAE but reduced with temperature, as well as IE got impressed insignificantly with temperature. The maximum value of IE was obtained at 94% at 4 gl^{-1} of the mentioned inhibitor. The Langmuir isotherm was suggested for adsorption of CAE on carbon steel.

Summary of the information in this section appears in Table 4.2.

4.3   IMPRESSION OF PLANTS IN HCL AND H$_2$SO$_4$ SOLUTION

Soltani and Khayatkashani [4] used weight loss, potentiodynamic polarization, and EIS methods to study the inhibitive function of leaf extract of Gundelia tournefortii (G. tournefortii) on mild steel in 2.0 M HCl and 1.0 M H$_2$SO$_4$ solutions. IE enhanced with the banner concentration. 93% and 90% IE at 150 ppm inhibitor in 2.0 M HCl and 1.0 M H$_2$SO$_4$ solutions was detected in the order. G. tournefortii was a mixed inhibitor in both acidic media. EIS information revealed that the charge transfer controls the corrosion. The addition of the mentioned banner extract in both acid solutions increased R$_{ct}$ values and decreased C$_{dl}$ values. IE decreased with a rise in temperature. Activation energies were higher in the presence of extract. More studies revealed that maybe protonated and neutral organic species in the extract were responsible for the observed inhibitive behavior, the major effect being the physical adsorption of protonated species. The adsorption of the inhibitor obeyed the Langmuir isotherm.

Table 4.2: Effect of plants in H_2SO_4 media

NO	Metal	Inhibitor	Additives	Methods	Findings	Ref
43	Mild steel	Musa Paradisiaca		Weight loss, potentiodynamic polarization, EIS	Mixed kind inhibitor, increase of IE with concentration of barrier and reduction of that by rising temperature, chemical and physical adsorption, spontaneous and exothermic adsorption, Langmuir isotherm	[111]
44	Mild steel	Solanum nigrum		Potentiodynamic polarization, AC impedance, SEM	Mixed kind, solasodine active compound, increase of IE with temperature, chemisorption mechanism following Temkin isotherm,	[112]
45	Mild steel	Ethanol extract of Gongronema Latifolium		Thermodynamic and gasometry	Physical adsorption mechanism and Langmuir isotherm	[113]
46	Steel	Artemisia		Electrochemical polarization, and linear polarization in the temperature range 298–353 K	Mixed inhibitor without changing the mechanism of the hydrogen evolution reaction, increase of IE with Artemisia concentration and temperature to attain 95% as well as 99% at 10 gl^{-1} at 298 and 353K, respectively, Langmuir isotherm	[114]
47	Carbon steel	Chenopodium Ambrosioides extract (CAE)		Weight loss, potentiodynamic polarization, and EIS	Cathodic inhibitor, reduction of IE with temperature, the maximum value of IE: 94% at 4 gl^{-1} of mentioned inhibitor, Langmuir isotherm	[115]

Oguzie et al. [55] examined Dacryodis edulis (DE) extract as a corrosion inhibitor on low carbon steel. Measurements were done at low and high concentrations (200 and 800 $\frac{mg}{li}$) of inhibitor in 1 M HCl and 0.5 M H_2SO_4 gravimetric and electrochemical methods, including potentiodynamic polarization and EIS were used to search corrosion rate and inhibition efficiency of inhibitor. Inhibition performance was more obvious by growth of DE concentration. Explorations showed greater inhibiting effect in 0.5 M H_2SO_4 than in 1 M HCl because of passivation at low concentration and the presence of ascorbic acid at high concentration by making $[Fe-AA]^{2+}$ and protecting the metal. DE acted as a mixed inhibitor. Adsorption behavior was approached to the Langmuir isotherm.

Studies represented physical adsorption of inhibitor species onto a corroding metal surface in 0.5 M H_2SO_4 because by rising in temperature, inhibition efficiency reduced. A similar behavior in 1 M HCl was seen at low concentration of DE, but at higher concentration inhibition efficiency increased by rising temperature which means chemical adsorption of inhibitor on metal.

Chlorophytum borivilianum root extract (CBRE) was tested to explore its corrosion inhibition ability for mild steel in 1 M HCl and 0.5 M H_2SO_4 solutions by means of weight loss, tafel polarization, EIS, and SEM. Prakash et al. [59] concluded CBRE functioned very well in both acidic solutions, however, a more significant effect was obtained in an HCl medium. Inhibition efficiency of CBRE depends on its concentration. IE increased by a rise in barrier concentration and decreased with an increase in immersion time, temperature of experiment, and acid concentration. Polarization studies exhibited mixed-type inhibition of barrier in both acid media affecting both anodic and cathodic reactions. Impedance data revealed inhibition action is due to chemical adsorption of banner substances which led to an increase in R_{ct} values and decrease in C_{dl} amounts. The Langmuir isotherm model was suggested as the most appropriate adsorption isotherm in both acid solutions.

The inhibitive effect of some plant extract was investigated in 1 M H_2SO_4 and 2 M HCl on mild steel by Oguzie [70]. Plants include leaf extracts of Occimum Viridis (OV), Telferia Occidentalis (TO), Azadirachta Indica (AI), and Hibiscus Sabdarifa (HS), as well as extracts from the seeds of Garcinia Kola (GK). It applied gasometric techniques at temperatures of 30 and 60°C.

Studies discovered that some plants banned corrosion in both acidic environments. Adsorption and inhibition efficiency of inhibitors got better with concentration. Protonated and neutral species had roles to prevent the action of corrosion. AI adsorbed physically on the mild steel at low concentration, but chemisorption was desired at high concentration. HS had chemical adsorption in 1 M H_2SO_4 and physical adsorption in 2 M HCl. TO and GK showed a major effect of physisorption. Inhibition efficiency in 1 M H_2SO_4 was, respectively: OV (70.2%) < GK (91.0%) < HS (93.0%) < AI (93.8%) < TO (97.3%), so except OV others had more than 90% efficiency.

Oguzie et al. [71] applied extract of punica granatum (PNG) to explore corrosion inhibition and the adsorption behavior on mild steel in 0.5 M H_2SO_4 and 1 M HCl. They utilized gravimetric, potentiodynamic polarization, and EIS methods. PNG functioned as a mixed inhibitor in both corrosive environments, but with distinguished effect on anodic reactions at high concentration in 0.5 M H_2SO_4 and prominent effect on cathodic reactions at low and high concentration in 1 M HCl. The adsorption of inhibitor followed Langmuir isotherm. Inhibition efficiency enhanced with increasing concentration of inhibitor and decreasing immersion time.

Deng and Xianghong [41] reported the inhibition effect of Ginkgo leaves extract (GLE) on the corrosion of cold rolled steel in 1.0–5.0 M HCl and 0.5–2.5 M H_2SO_4 solutions. They used weight loss, potentiodynamic polarization, and electrochemical impedance spectroscopy. GLE was a good inhibitor and a mixed banner in 1 M HCl while a cathodic in 0.5 M H_2SO_4 and was more effective in 1 M HCl as well as in 0.5 M H_2SO_4. The adsorption behavior obeyed the Langmuir isotherm. Inhibition efficiency in 1 M HCl was more than 0.5 M H_2SO_4. EIS data showed that with the addition of GLE in both acid solutions R_{ct} values increase while C_{dl} values diminish.

Behpour et al. [62] studied the effect of Punica granatum (PG) extract and its major substances such as ellagicacid (EA) and tannic acid (TA) on the corrosion of mild steel in 2.0 M HCl and 1.0 M H_2SO_4. It utilized the weight loss method which exhibited IE (inhibition efficiency) of TA even in high concentration is very slow. PG and EA acted as mixed inhibitors. Inhibition performance of PG peel extract and EA rarely has been better in 2.0 M HCl than in 1.0 M H_2SO_4 solution. Transfer resistance increased with an increase of banner concentration. The adsorption of both PG peel and EA followed the Langmuir isotherm.

The study on corrosion inhibition of mild steel using an aqueous extract of Allamonda Blanchetti (AB) as natural inhibitor in 2 M H_2SO_4 and 2 M HCl environment at room temperature to control the rate of corrosion was carried out by Anand and Balasubramanian [116]. It applied the weight loss method and adsorption isotherm. Results showed mentioned that the inhibitor was effective. IE enhanced by rising banner concentration. It was concluded that IE was better in 2 M H_2SO_4 than HCl. The adsorption of banner molecules was physical and followed the Langmuir isotherm.

Green tea as an environmentally friendly corrosion inhibitor for carbon steel in both 1 M HCl and 0.5 M H_2SO_4 solutions was evaluated by Khamis et al. [117] applying potentiodynamic polarization and EIS methods. According to the results, banner acted as a mixed type. IE increased with a rise in concentration of extract while the corrosion rate reduced in both acidic media. The maximum IE was achieved 81.47% in 1 M HCl and 71.65% in 0.5 M H_2SO_4 at concentration of 500 ppm. The adsorption of banner followed the Langmuir isotherm. R_{ct} increased and C_{dl} decreased by increasing banner concentration.

The adsorption and corrosion inhibiting effect of acid extracts of piper guineense (PG) leaves on mild steel corrosion in 0.5 M H_2SO_4 and 1 M HCl was studied by Oguzie et al. [118] using gravimetric, potentiodynamic polarization, and EIS methods, as well as SEM and FT-

IR. PG inhibited corrosion of mild steel by influencing anodic and cathodic reactions, so it played as a mixed-type barrier. Chemisorptive adsorption of extract substances and the Langmuir isotherm were suggested for the adsorption of active elements of inhibitors on a metal surface.

Summary of the information in this section appears in Table 4.3.

4.4   INFLUENCE OF PLANTS IN NACL SOLUTION

Hadi [119] explored the inhibitive properties of natural products like reed leaves as a corrosion inhibitor for low carbon steel in 3.5% NaCl solution by using the weight loss method. He discovered the amount of weight loss of low carbon was diminished by increasing banner concentration. Reed leaves acted as a safety and an environmentally friendly inhibitor for low carbon in aqueous media.

Tafel extra polarization, electrochemical impedance spectroscopy, and dynamic electrochemical impedance spectroscopy were applied by Gerengi et al. [69] to evaluate inhibition efficiency of mad honey on corrosion of 2007-type Al alloy in 3.5% NaCl solution. It was found that mad honey functioned as a mixed-type inhibitor. IE increased by the growth in banner concentration. Mad honey adsorbed on the surface of Al-2007 alloy physically and the Langmuir isotherm was offered for the adsorption of mentioned inhibitor on Al surface.

El-Etre [120] investigated the inhibition action of natural honey on the corrosion of copper in 0.5 M sodium chloride solution using weight loss and cathodic polarization. It was found that natural honey performed as a good inhibitor and shifted the corrosion potential toward the cathodic direction. IE increased by a rise in honey concentration. After some days, because of the growth of microorganisms in the examined medium, IE decreased. The adsorption of mentioned inhibitor obeyed the Langmuir isotherm.

The application of Momordica Charantia seeds extract (MCSE) as an environmentally benign corrosion inhibitor for J55 steel 3.5% Nacl solution was evaluated by Lin et al. [121]. It utilized polarization carve, AC impedance, and X-ray diffraction (XRD). MCSE worked as a mixed-type inhibitor. EIS information showed an increase in R_{ct} values which means good inhibition efficiency. The highest inhibition efficiency was obtained when MCSE concentration was 1,000 ppm by weight. The adsorption of inhibitor molecules on the mentioned metal surface obeyed the Langmuir isotherm.

Lin et al. [122] studied Root extract of coptis chinensis (CCR) as a natural inhibitor for corrosion inhibition of Al alloy at 3.5% NaCl solution by surface morphology and electrochemical studies. CCR acted as a mixed-type inhibitor. An increase in R_{ct} values was detected which means a rise in inhibition efficiency. The adsorption of inhibitor on the Al alloy surface followed the Temkin isotherm.

Al-Asadi et al. [123] studied the effect of a new natural inhibitor distinguished as an aloe vera (mannose-6-phosphate) on the corrosion of mild steel in 1 wt% NaCl by applying potentiodynamic polarization measurement. This compound acted as an effective inhibitor. IE

Table 4.3: Impression of plants in HCl and H_2SO_4 solution (*Continues*)

NO	Metal	Inhibitor	Additives	Methods	Findings	Ref
48	Mild steel	Leaf extract of Gundelia tournefortii (G. tournefortii)		Weight loss, potentiodynamic polarization, EIS	Mixed inhibitor, decrease in IE with rise of temperature, Langmuir isotherm	[4]
49	Low carbon steel	Dacryodis edulis (DE)		Gravimetric and electrochemical methods such as potentiodynamic polarization and EIS	Mixed-type inhibitor, greater IE in H_2SO_4, Langmuir isotherm.	[55]
50	Mild steel	Chlorophytum borivilianum root (CBR)		Weight loss, tafel polarization, EIS, SEM	Mixed-type inhibitor, more significant effect in HCl medium, reduction of IE with increase in immersion time, temperature of experiment, and acid concentration, Langmuir isotherm, chemical adsorption	[59]
51	Mild steel	Leaf extracts of Occimum Viridis (OV), Telferia Occidentalis (TO), Azadirachta Indica (AI), Hibiscus Sabdarifa (HS), and extracts from the seeds of Garcinia Kola (GK)		Gasometry at 30 and 60°C	Improvement of IE by a rise of inhibitor concentration, the order of inhibitive of banners: OV (70.2%)<GK (91.0%)<HS (93.0%)<AI (93.8%)<TO (97.3%)	[70]
52	Mild steel	Punica granatum (PNG)		Weight loss, potentiodynamic polarization, EIS	Mixed-type inhibitor, Langmuir isotherm, enhancement of IE with rise of PNG concentration and reduction of immersion time	[71]
53	Cold rolled steel	Ginkgo leaves (GL)		Weight loss, potentiodynamic polarization, EIS, SEM	Mixed-type and cathodic inhibitor in 1 M HCl and 0.5 M H_2SO_4, respectively, more IE in 1 M HCl than 0.5 M H_2SO_4, Langmuir isotherm	[41]

Table 4.3: (*Continued*) Impression of plants in HCl and H$_2$SO$_4$ solution

54	Mild steel	Punica granatum (PG) extract and their major substances such as ellagicacid (EA) and tannicacid (TA)		Weight loss, potentiodynamic polarization, EIS, SEM	Mixed mode inhibitor for PG and EA, slowness of inhibition performance of TA even at high concentration, prevention of corrosion by PG peel extract and EA in two acidic environments, Langmuir isotherm	[62]
55	Mild steel	Allamonda Blanchetti (AB)		Weight loss method, adsorption isotherm, FT-IR, at room temperature	Better IE in 2 M H$_2$SO$_4$ than HCl, physical adsorption, Langmuir isotherm	[116]
56	Carbon steel	Green tea		Potentiodynamic polarization, EIS, SEM	Mixed type, the maximum of IE in 1 M HCl and 0.5 M H$_2$SO$_4$ at concentration of 500 ppm: 81.47%, 71.65%, respectively, Langmuir isotherm	[117]
57	Mild steel	Acid extracts of piper guineense (PG) leaves		Gravimetric, potentiodynamic polarization, and EIS methods, as well as SEM and FT-IR	Mixed-type barrier, chemisorptive adsorption, Langmuir isotherm	[118]

and coverage ratio increased by rising in aloe vera concentration to reach 81.81% and 0.818%, respectively. By increasing the concentration of aloe vera above 75 μml, IE and coverage rate decreased. Aloe vera performed as a mixed kind of banner for mild steel in a NaCl solution. The adsorption of an inhibitor was mainly physical.

The inhibition effect of santolina chamaecyparissus extract as a natural inhibitor on corrosion of 304 stainless steel in 3.5% Nacl solution was tested by Shabani-Nooshabadi and Ghandchi [124] utilizing potentiodynamic polarization, EIS, and SEM techniques. It was concluded that the inhibitor functioned as a mixed-type inhibitor because addition of mentioned banner in concentration from 0.2–1.0 gL^{-1} predominately causes a decrease in the cathodic and anodic currents. IE enhanced by a rising in concentration of an inhibitor and decreased by rising temperature. Corrosion rate became 86.9% for 1 gL^{-1} of extract. Charge transfer increased as the concentration of extract went up, and double-layer capacity reduced. Active molecules from inhibitor postponed the corrosion on specimen surface which was concluded by SEM technique. The adsorption of extract obeyed the Langmuir isotherm.

Analysis of mechanism of Morinda lucida leaf extract blend on concrete steel Rebar immersed in 3.5% NaCl to simulate saline/Marine environment was carried out by Okeniyi et al. [125]. According to the results, 3.333 gl^{-1} Morinda lucida showed optimal inhibition efficiency, IE: 90.59±2.52 (by correlation prediction) or IE: 89.27±3.94 (by experimental model). The adsorption of extract was physical and followed the Langmuir isotherm.

Abdullatef et al. [126] investigated inhibitive influence of Lupine, Hlfabar, and Damssesa on corrosion of zinc in 0.5 M NaCl solution. Potentiodynamic polarization and EIS techniques were used. Three inhibitors acted as mixed type. The maximum values of IE reached to 89.1%, 94.7%, and 90.7% at 40, 40, and 15 ppm for Lupine, Hlfabar, and Damssesa extract, respectively. IE reduced at higher concentration. IE of three plant extracts was in the order: Damssesa > Hlfabar > Lupine.

The effect of natural honey (chestnut and acacia) in 3% NaCl solution as well as their solutions with black radish juice added, play a role as corrosion inhibitors of tin, was studied by Kovac et al. [127] with the help of weight loss and polarization techniques. Based on the acquired results, inhibition action was ascribed to the formation of a multilayer adsorbed film on the tin surface. Inhibition performance of all inhibitors reduced in the order: chestnut honey with black radish juice > acacia honey with black radish juice > chestnut honey > acacia honey. IE of honey and honey with black radish juice enhanced with rise of inhibitor concentration. The Langmuir isotherm was described for the adsorption of natural honey and honey with black radish on tin.

Protection effect of Juniperus Communis (JC) extract on the corrosion of synthetic bronze (cu10 Sn) in natural 0.5 M NaCl solution was studied by Benchannouf et al. [128] who applied electrochemical polarization methods. It was found the maximum value of inhibition in chloride solution at 313 K and 15% concentration of extract was achieved 75%. The adsorption of JC on bronze was physical, endothermic, and followed Arrhenius isotherm, as well as reduced the molecular disorder in the interface.

Summary of the information in this section appears in Table 4.4.

4.5 EFFECT OF PLANTS IN OTHER CORROSIVE ENVIRONMENTS

El-Etre and Abdallah [129] studied the inhibition action of natural honey as an inhibitor for C-steel A 106 in high saline water. It used weight loss and potentiostatic polarization methods. Natural honey functioned as a good inhibitor. In high saline water, IE increased with inhibitor concentration. After a while, IE reduced thanks to the growth of fungi in the medium. The adsorption of natural honey on the C-steel in high saline solution followed the Langmuir isotherm.

Commercial henna (Lawsonia inermis) was studied as an inhibitor for aluminum in sea water by Nik et al. [130]. It applied weight loss, Fourier Transform Infrared (FT-IR), and EIS. Henna performed as an excellent inhibitor. The inhibition action of henna is related to the main substances of lawsone which contributed to the chemisorptions. The adsorption of inhibitor on the metal surface followed Langmuir isotherm. The highest inhibition efficiency was detected in 88% (500 ppm).

The efficiency of croton cajucara Benth (CC) dissolved in a microemolusion system (MES-CC) as well as in dimethyl sulfoxide (DMSO-CC) on corrosion inhibition of carbon steel in saline medium was tested by Maciel et al. [131]. Potentiodynamic method and tafel extrapolation showed maximum and impressive inhibition efficiency (93.84% for MES-CC and 64.63% for DMSO-CC) with main control of cathodic reaction. The adsorption of MES-CC on carbon steel surface followed Langmuir isotherm while DMSO-CC was discovered to obey the Frumkin isotherm.

Ayende et al. [132] found out some plants including Manikara Zapota or sawo, Garcinia Mangostana L or mangosteen, and Ipomea Batatas or purple sweet potato have inhibition effect on API-5L in produce water using polarization curves. Conclusion of this research revealed stated banners can reduce corrosion rate, however, they weren't as potent as chemical inhibitors. Sawo and purple sweet potato were classified as anodic inhibitors while mangosteen was classified to cathodic inhibitor. All of inhibitors exhibited instability.

Mhinzi and Buchweishaija [133] investigated the inhibitive effect of the gum exudate on the corrosion of mild steel in drinking water utilizing potentiodynamic polarization and EIS methods. Based on conclusions, gum behaved as an anodic inhibitor. Percentage of IE increased above 95% at 30°C at gum concentration $\geq$ 400 ppm. The study also exhibited temperature didn't influence the adsorption of the gum. Chemical mechanism was proposed for the adsorption of mentioned gum on mild steel.

Corrosion inhibition of st37 in Geothermal fluid by Quercus robur and pomegranate peels extracts in aqueous media was tested by Buyuksagis et al. [134] using the tafel polarization

Table 4.4: Influence of plants in NaCl solution (*Continues*)

NO	Metal	Inhibitor	Additives	Methods	Findings	Ref
58	Low carbon steel	Reed leaves		Weight loss	Reduction of amount of weight loss of low carbon with rise of inhibitor concentration	[119]
59	2007-type Al alloy	Mad Honey		Tafel extra polarization, electrochemical impedance spectroscopy, and dynamic electrochemical impedance spectroscopy	Mixed-type inhibitor, physical adsorption, Langmuir isotherm	[69]
60	Copper	Natural honey		Weight loss, cathodic polarization	Change corrosion potential toward cathodic direction, Langmuir isotherm,	[120]
61	J55 steel	Momordica Charantia seeds (MCS)		Polarization carve, AC impedance, X-ray diffraction (XRD)	Mixed mode inhibitor, Langmuir isotherm	[121]
62	Al alloy	Root extract of coptis chinensis (CCR)		Potentiodynamic polarization, EIS, SEM, UV-visible spectroscopy	Mixed-type inhibitor, Temkin isotherm	[122]
63	Mild steel	Aloe vera (mannose-6-phosphate)		Potentiodynamic polarization	Mixed mode inhibitor, the maximum of IE: 81.81%	[123]
64	304 stainless	Santolina chamaecyparissus		Potentiodynamic polarization, EIS, SEM	Mixed mode banner, increase of IE with rise of barrier concentration and reduction of that with rise of temperature, Langmuir isotherm	[124]
65	Concrete steel Rebar	Morinda lucida leaf		Electrochemical tests	Physical adsorption, Langmuir isotherm	[125]
66	Zinc	Lupine, Hlfabar and Damssesa		Potentiodynamic polarization, EIS	Mixed-type inhibitor, Inhibition efficiency of three plant extracts in the order: Damssesa> Hlfabar> Lupine	[126]

Table 4.4: (*Continued*) Influence of plants in NaCl solution

67	Tin	Honey (chestnut and acacia), black radish juice		Weight loss, polarization	IE of all inhibitors: chestnut honey with black radish juice > acacia honey with black radish juice > chestnut honey > acacia honey, Langmuir isotherm	[127]
68	Synthetic bronze (cu10 Sn)	Juniperus Communis (JC)		Electrochemical polarization	75% the maximum value of inhibition in chloride solution at 313 K and 15% concentration of extract, physical,and enthothermic adsorption, Arrhenius isotherm	[128]

method. Extracts played as good and mixed inhibitors. 250 mgl^{-1} extract of Quercus robur oak and 500 mgl^{-1} extract of pomegranate peel indicated 90% inhibition efficiency.

According to Ryznar and Langelier indexes calculations geothermal fluid has corrosive feature and is scale properties. The scale formation happened very little because there was inhibitor in geothermal fluid sample taken for index computation which inhibits scale formation but it doesn't prevent corrosion. So both corrosion inhibitor and scale inhibitor should be added into the system to ban corrosion process and scale formation.

Inhibition of cu65/zn35 brass corrosion was studied by Ramde et al. [135] using camellia sinensis as a natural inhibitor in 1 M Na$_2$SO$_4$ solutions with pH 7 and pH 4. Electrochemical techniques including potentiodynamic polarization, electrochemical impedance spectroscopy, and SEM were applied. According to obtained conclusions, camellia sinensis extract revealed effective inhibition for brass in both acidic and neutral 0.1 M Na$_2$SO$_4$; it also played as a mixed banner. Inhibition efficiency enhanced with time. In the absence of inhibitor a dark oxide patina was formed at pH 7 because of corrosion process and caused localize corrosion morphology at pH 4, but the presence of the inhibitor inhibited both dark patina and pits formation.

Improvement of corrosion performance of 316L stainless steel via hybrid sol-gel coating was studied by Nasr-Esfahani et al. [136]. An ethanol solution of the polymerized vinyltrimethoxysaline (PVTMS) was combined with an aqueous solution of henna extract (law Sonia inermis) and refluxed in order to give alike sols. Nano structure hybrid PVTMS/henna thin layers were deposited on the stainless steel 316L by spin-coating. The anti-corrosive properties of sol gel which is attributed to henna extract was examined by EIS, potentiodynamic polarization tests resembling body fluid solution. Obtained results indicated the addition of 50% henna predominantly enhanced corrosion protection of sol-gel thin films to higher than

90%. Henna extract played as a mixed-type barrier, mainly cathodic. It was found that doped PVTMS thin film on the stainless steel 316L can be suggested as bioactive thin film for biomedical applications.

Effect of lupine and damsissa extracts on the corrosion of mild steel in 0.5 M Na_2SO_4 free from and including 0.01 or 0.1 M NaCl was evaluated by Abdel-Gaber et al. [137] using potentiodynamic and EIS methods. Lupine and damsissa were found to be anodic kind inhibitor. IE increased by rising chloride ion concentration due to co-operative mechanism of inhibition. Lupine extract performed more effective than damsissa extract.

Abdel-Gaber et al. [63] utilized damsissa (Ambrosia maritime, L.) extract to investigate corrosion of aluminium in 2 M NaOH solution in the presence and absence of 0.5 M NaCl. Chemical and electrochemical measurements including gasometry, potentiodynamic polarization, and EIS methods were exerted. It was concluded that Damsissa extract functioned as an impressive inhibitor for the alkaline corrosion of aluminium. The mentioned inhibitor monitored dissolution of AL in alkaline media in the presence and absence of chloride ions. The presence of chloride ions hindered anodic dissolution of Al. An inhibitive mechanism was independent on the storage time.

The extract of lawsonia, licorice root, and corob were used as corrosion inhibitor of tin electrode in 0.1 M $NaHCO_3$ by Abdallah et al. [138] applying galvanostatic polarization measurements. All of mentioned extracts acted as mixed inhibitors. Inhibition performance of these extracts is related to the adsorption of its substances on tin surface. Spontaneous process and freundlich isotherm were suggested for the adsorption of inhibitors on tin surface. These inhibitors hindered the pitting corrosion of tin in chloride including solutions.

Torres-Acosta [139] investigated inhibition performance of dehydrated opuntia ficus indica (Nopal) mixed with saturated calcium hydroxide on the corrosion of reinforcing steel using half cell potentials and linear polarization resistance. According to the results, the addition of dehydrated Nopal to saturated calcium hydroxide solution in less than 2% increased corrosion resistance. Polarization resistance values of reinforcing steel in such alkaline solution with addition of Nopal were higher than values for the same steel in alkaline solution without Nopal (without Nopal ~40 KΩ cm^2 vs. with Nopal ~50 KΩ m^2 in average). The addition of Nopal caused to form a layer of oxide/hydroxide on the steel surface which reduced corrosion action.

Summary of the information in this section appears in Table 4.5.

Based on plant parts, we can conclude the use of inhibitors is one of the most practical methods for protection of metals in corrosive media. However, most organic inhibitors are harmful and poisonous to the environment. This has led to the need for natural products which are eco-friendly and harmless.

Some investigations have reported the use of natural banners which were extracted from plants extracts. Such studies are justified by phytochemical compounds present in plants with molecular and electronic structures bearing close similarity to those conventional organic inhibitor molecules.

Table 4.5: Effects of plants in other corrosive environments (*Continues*)

NO	Metal	Medium	Inhibitor	Additives	Methods	Findings	Ref
69	C-steel A 106	High saline water	Natural honey		Weight loss, potentiostatic polarization	Langmuir isotherm	[129]
70	Aluminum (Al)	Sea water	Commercial henna (Lawsonia inermis)		Weight loss, Fourier Transform Infrared (FT-IR), EIS	Lawsone the main active component, chemical adsorption, maximum of IE: 88%, Langmuir isotherm	[130]
71	Carbon steel	Saline medium	Croton cajucara Benth (CC) dissolved in a microemolusion system (MES-CC) as well as in dimethyl sulfoxide (DMSO-CC)		Potentiodynamic and Tafel extrapolation	Impressive inhibition efficiency (93.84% for MES-CC and 64.63% for DMSO-CC) with main control of cathodic reaction, Langmuir isotherm for MES-CC, Frumkin isotherm for the adsorption of DMSO-CC	[131]
72	API-5L	Produce water	Manikara Zapota or sawo, Garcinia Mangostana L or mangosteen, and Ipomea Batatas or purple sweet potato		Polarization curves	Sawo and purple sweet potato, anodic inhibitors, mangosteen, cathodic inhibitor,	[132]
73	Mild steel	Drinking water	Gum exudate		Potentiodynamic polarization, EIS	Anodic inhibitor, increase of IE above 95% at 30°C at gum concentration ≥ 400 ppm, temperature didn't influence the adsorption of the gum, chemical adsorption mechanism	[133]

Table 4.5: (*Continued*) Effects of plants in other corrosive environments (*Continues*)

74	St37	Geother-mal fluid	Quercus robur and pomegranate peels		Tafel polarization	Mixed inhibitors, 90% inhibition efficiency with 250 mgl^{-1} extract of Quercus robur oak and 500 mgl^{-1} extract of pomegranate peel	[134]
75	cu65/zn35 brass	1 M Na$_2$SO$_4$ solutions with pH 7 and pH 4	Camellia sinensis		Potentiodynamic polarization, electrochemical impedance spectroscopy, SEM	Mixed banner, increase of IE by time, inhibition of both dark patina and pits formation by using inhibitor	[135]
76	316L stainless steel	An ethanol solution of the polymerized vinyltri-methox-ysaline (PVTMS)	Henna extract (law Sonia inermis)		EIS, potentiodynamic polarization	Mixed-type barrier mainly cathodic, higher than 90% at 50% henna	[136]
77	Mild steel	0.5 M Na$_2$SO$_4$	Lupine and damsissa	0.01 or 0.1 M NaCl	Potentiodynamic, EIS	Anodic kind inhibitor, better performance of IE for lupine than damsissa	[137]
78	Aluminium	2 M NaOH	Damsissa (Ambrosia maritime, L.)	0.5 M NaCl	Gasometry, potentiodynamic polarization, EIS	Control of dissolution of Al in alkaline media in the presence and absence of chloride ions by inhibitor, independence of inhibitive mechanism on the storage time	[63]
79	Tin electrode	0.1 M NaHCO$_3$	Lawsonia, licorice root, and corob	Chloride	Galvanostatic polarization	Mixed inhibitors, spontaneous process and freundlich isotherm, to hinder pitting corrosion of tin in chloride including solutions	[138]

Table 4.5: (*Continued*) Effects of plants in other corrosive environments

| 80 | Reinforcing steel | Alkaline media | Dehydrated opuntia ficus indica (Nopal) mixed with saturated calcium hydroxide | | Half-cell potentials and linear polarization resistance | Higher polarization resistance values of reinforcing steel in alkaline solution with addition of Nopal than values for the same steel in alkaline solution without Nopal | [139] |

In addition, plants are low-cost, readily available, and renewable sources of materials. Despite these desirable features, only relatively few have been thoroughly investigated, and even at that reports detailed mechanism of the adsorption process are still scarce.

Moreover, plant materials containing phenolic ingredients are becoming of great interest to many investigations thanks to their anti-oxidative activities and nutritional value.

CHAPTER 5

Fruits as Corrosion Inhibitors in Corrosive Environments

5.1 EFFECT OF FRUITS IN HCL MEDIA

Gomes et al. [5] utilized weight loss, potentiodynamic polarization, and EIS measurements as well as surface analysis to investigate inhibition effect of mango, orange, passion fruit, and cashew peels on corrosion of carbon steel in 1 M HCl. IE increased with the rise of inhibitor concentrations and diminished with the increase of temperature. IE was achieved in the presence of $400\frac{mg}{l}$ extracts in the order: cashew < mango < passion fruit < orange. The adsorption of banners obeyed Langmuir isotherm.

Opuntia ficus-indica (Nopal) was evaluated as green corrosion inhibitor for carbon steel in 1 M HCl solution by Flores-De los Rois et al. [140]. Weight loss tests, potentiodynamic polarization curves, EIS, SEM, and FT-IR analysis were studied. At room temperature, IE was highest at 75 ppm, but at higher temperatures IE enhanced by increasing extract concentration and reduced with rising temperature. Inhibitor functioned as a cathodic type which was adsorbed spontaneously and physically. Adsorption of banner on the steel surface obeyed Langmuir isotherm showing monolayer adsorption.

Investigation of Diospyros Kaki L.f (persimmon) husk extract, a famous fruit tree, as a corrosion inhibitor of Q235A steel in 1 M HCl solution and bactericide in oil field was carried out by Chen et al. [141] using weight loss and potentiodynamic polarization methods. The water and alcohol extract of persimmon husk named as WE and AE, respectively, exhibited temperate to high impressive inhibition in the range of 10–1,000 ppm in 1 M HCl at 60°C and the maximum protection 65.1% was achieved by applying WE solution of 1,000 ppm. IE also enhanced to 97.3% at most by addition of KI, SCN, and HMTA (Hexamethylene tetramine) as synergistic additives, but not impressive for KI and SCN to AE. The mentioned barrier was indicated to be as mixed type. Exploration of antibacterial against oil field micro organism indicated, SRB (sulfate reducing bacteria), IB (iron bacteria), and TGB (total general bacteria) were inhibited by extracts from temperature to highly performance under 1,000 ppm which makes extracts potential to be utilized as bifunctional oil field chemicals.

Skin extracts of some fruits including mango, cashew, passion-fruit, and orange were used as corrosion inhibitors for carbon steel 1020 in 1 M HCl by Gomes [142]. Electrochemical treatment of carbon steel was explored by electrochemical impedance measurements, anodic polarization, cathodic curves, and mass loss tests at room temperature for each solution. It was concluded that the peels of fruits acted as good nature inhibitors. Notable inhibition occurred in both cathodic and anodic reactions in the presence of all extracts by reducing current density. The best result of IE was obtained for the extract of orange peel (IE: 95% at 400 ppm) and the lowest amount of IE was achieved for the peel of cashew (IE: 80% at 800 ppm). Polarization resistance increased by concentration of all mentioned inhibitors. The main ingredients of inhibitors include carotenoids and phenolic compounds from cashew, flavanoids, alkaloids, and pectin from passion fruit, flavanoids, carotenoids, and pectin from orange, polyphenols, carotenoids, enzymes, and fiber from mango.

Table 5.1 shows the effects of fruits in HCl media.

Table 5.1: Effects of fruits in HCl media

NO	Metal	Inhibitor	Additives	Methods	Findings	Ref
81	Carbon steel	Mango, orange, passion fruit, and cashew peels		Weight loss, potentiodynamic polarization, EIS, surface analysis	Reduction of IE with increase of temperature, inhibition efficiency of banners: cashew < mango < passion fruit < orange, Langmuir isotherm	[5]
82	Carbon steel	Opuntia ficus-indica (Nopal)		Weight loss tests, potentiodynamic polarization curves, EIS, SEM, FT-IR	Cathodic-type inhibitor, spontaneous and physical adsorption, Langmuir isotherm,	[140]
83	Q235A steel	Diospyros Kaki L.f (persimmon) husk extract	KI, SCN, HMTA (Hexamethylene tetramine)	Weight loss, potentiodynamic polarization	Mixed inhibitor, enhancement of IE to 97.3% at most by addition of KI, SCN, and HMTA (hexamethylene tetramine) as synergistic additives	[141]
84	Carbon steel 1020	Skin extracts of mango, cashew, passion fruit, orange		Mass loss, anodic polarization, cathodic curves, EIS	Appearing notable inhibition in both cathodic and anodic reactions, obtaining the best result of IE by orange peel and the lowest by cashew peel	[142]

5.2   EFFECT OF FRUITS IN H$_2$SO$_4$ AND HCL SOLUTION

Amudha et al. [143] investigated the efficiency of L. Acidissima as a natural inhibitor on corrosion of mild steel in 1 M and 2 M H$_2$SO$_4$ at room temperature. Weight loss measurements and FT-IR spectroscopic methods were applied. Studies showed that the mentioned inhibitor acted well in 1 M H$_2$SO$_4$ solution at room temperature in 2 h. IE was better in 1 M H$_2$SO$_4$ than 2 M H$_2$SO$_4$ solution. Inhibition efficiency through electrostatic adsorption of banner molecule on metal surface was verified.

Mahdi [144] studied the pomegranate peel powder as corrosion inhibitor for mild steel in 5% HCl and 5% H$_2$SO$_4$ solution by applying weight loss, electrochemical tafel test, and optical and SEM examination with EDX. Results revealed the pomegranate peel powder acted as a good inhibitor for mild steel in diluted strong acids for a short period. The pomegranate peel powder was insoluble in acid solutions, and its reactant products precipitate on a metal surface to form a protective layer on the surface of metal.

Potentiodynamic polarization, linear polarization, EIS, and SEM were used to investigate the inhibition effect of watermelon rind extract (WMRE) on corrosion of mild steel in 1 M HCl and 0.5 M H$_2$SO$_4$ at 25°C by Odewunmi et al. [145]. WMRE was found to be a more effective inhibitor in 1 M HCl than in 0.5 M H$_2$SO$_4$ and acted as a mixed inhibitor by impressing both anodic dissolution of mild steel and cathodic hydrogen evolution. IE increased by a rise in concentration of WMRE and the optimum concentration was 1.5 gl^{-1}. Physical adsorption and obeying Temkin isotherm was discovered by mentioned banner.

Corrosion inhibition of mild steel with Date Palm (phoenix dactylifera) seed extracts (DPSE) in acid media was explored by Umoren et al. [146]. Weight loss and electrochemical methods including linear polarization, potentiodynamic polarization, and EIS at 25 and 60°C were utilized. Based on obtained results Date Palm acted as a mixed inhibitor and hindered corrosion process in 1 M HCl and 0.5 M H$_2$SO$_4$ solutions. IE depends on concentration, 2.5 gL^{-1} and 1.5 gL^{-1} of extract gave the highest IE in 1 M HCl and 0.5 M H$_2$SO$_4$, respectively. Inhibition action became affected by immersion time. With an increase in immersion time and the extract concentration, IE increased in 1 M HCl and reduced in 0.5 M H$_2$SO$_4$ while the extract concentration increased up to 2.0 gL^{-1}. DPSE was found to be better inhibitor in 1 M HCl than in 0.5 M H$_2$SO$_4$. The adsorption of DPSE components on to mild steel was suggested to follow the Langmuir isotherm and physical adsorption for the mechanism of that.

The anticorrosive effect of opuntia elatior fruit extract was explored on mild steel (MS) in 1 M HCl and H$_2$SO$_4$ solution by Sethuraman et al. [147] with the help of weight loss, potentiodynamic polarization, EIS, and surface studies including SEM. It was found that IE increased with a rise of extract concentration and reduced with an increase in temperature which suggested physical adsorption. The mixed mode inhibitive was revealed by extract. The adsorption of inhibitor on the mild steel surface followed Temkin isotherm. The anticorrosive action of O. elatior was attributed to the presence of opuntiol. Based on surface results, a protective layer was made on mild steel by opuntiol. Of course other phytoconstituents such as proline, linolenic acid, campesterol, and betacyanin had inhibition properties.

Summary of the information in this section appears in Table 5.2.

5.3   IMPRESSION OF FRUITS IN OTHER CORROSIVE MEDIA

Khadom et al. [72] examined influences of apricot juice as a green corrosion inhibitor on mild steel in 1 M H_3PO_4 solution by utilizing weight loss method at different temperatures. An inhibitor was adsorbed by making a layer on the metal surface according to the Langmuir isotherm. A spontaneous physical adsorption on the metal surface was shown. The maximum inhibition efficiency at the greatest level of inhibitor concentration and at the 30°C was 75%. Statistical analysis revealed that corrosion rate depends on temperature, inhibitor concentration, and the combined effect of them.

Summary of the information in this section appears in Table 5.3.

Another group of natural inhibitors includes fruits which are one of the best options for protecting metals and alloys against corrosion. They are biodegradable and don't contain heavy metals. These features make them appropriate compared with organic and inorganic banners.

Fruit is a rich source of chemicals such as vitamins, minerals, and phenolic compounds which can protect metals by adsorbing on their surface and blocking active sites for metal dissolution and/or hydrogen evolution, thereby hindering overall metal corrosion in aggressive media.

Table 5.2: Effect of fruits in H_2SO_4 and HCl solution

NO	Metal	Inhibitor	Additives	Methods	Findings	Ref
85	Mild steel	L. Acidissima		Weight loss, FT-IR spectroscopic, SEM	Better IE in 1 M H_2SO_4 than 2 M H_2SO_4	[143]
86	Mild steel	Pomegranate peel powder		Weight loss, electrochemical tafel test, optical and SEM examination with EDX	Protection of metal surface by precipitating inhibitor reactant products to form a protective layer	[144]
87	Mild steel	Watermelon rind (WMR)		Potentiodynamic polarization, linear polarization, EIS, SEM	Mixed inhibitor, more effectiveness of inhibitor in 1 M HCl than in 0.5 M H_2SO_4, physical adsorption, Temkin isotherm	[145]
88	Mild steel	Date Palm (phoenix dactylifera) seed (DPS)		Weight loss and electrochemical methods including linear polarization, potentiodynamic polarization, and EIS at 25 and 60°C	Mixed inhibitor, the highest IE in 1 M HCl and 0.5 M H_2SO_4 at the concentration of 2.5 g L^{-1} and 1.5 g L^{-1} of extract, increase of IE in 1 M HCl and reduction of that in 0.5 M H_2SO_4 with increase immersion time and the extract concentration, being a better inhibitor in 1 M HCl than in 0.5 M H_2SO_4, Langmuir isotherm and physical adsorption	[146]
89	Mild steel	Opuntia elatior		Weight loss, potentiodynamic polarization, EIS, and surface studies including SEM	Mixed mode inhibitor, reduction of IE with an increase in temperature, physical adsorption, Temkin isotherm,	[147]

Table 5.3: Impression of fruits in other corrosive media

NO	Metal	Medium	Inhibitor	Additives	Methods	Findings	Ref
90	Mild steel	1 M H_3PO_4	Apricot juice		Weight loss method at different temperature	Langmuir isotherm, physical and spontaneous adsorption, Max of IE:75%	[72]

CHAPTER 6

Natural Oils as Corrosion Inhibitors in Corrosive Environments

6.1 INFLUENCE OF NATURAL OILS IN HCL SOLUTION

Popoola [3] studied the electro-oxidation behavior and passivation potential of natural oils (Arachis hypogeae) as corrosion inhibitors for mild steel in 2 M HCl solution by using gravimetric, potentiodynamic polarization, and SEM-EDX. Arachis hypogeae functioned as a good corrosion banner for mild steel in HCl solution at 25°C. With addition of mentioned inhibitor, corrosion rate decreased and IE increased as high as 75.19 and 94.99% using gravimetric and electrochemical, respectively, in HCl-Arachis hypogeae condition. The mixed type corrosion inhibition and Langmuir adsorption isotherm were proposed for the mild steel.

Avogadro natural oil demonstrated a good banner for mild steel in 1 M HCl environment at 298 K using gravimetric and potentiodynamic polarization methods by Abduwahab et al. [148]. The addition of natural oil reduced corrosion rate. Inhibition efficiency increased by concentration of barrier and reached 81% at 405 $\frac{g}{V}$ of Avogadro natural oil. The Langmuir isotherm was suggested for the adsorption of a banner. Inhibition action was ascribed to the invention of thin oxides which stuck to metal surface and interfered with the reaction sites, thereby serving as a barrier to the formation of pits and their growth.

The behavior of corrosion of tinplate in 0.5 M HCl media was rated with the addition of Artemisia essential oil (AO) as a green inhibitor by Hammouti et al. [149] applying potentiodynamic polarization. It was found that AO worked as a mixed kind inhibitor and its performance was excellent. Corrosion rate of tinplate decreased with the addition of AO. Dissolution of tinplate in HCl media was motivated by increase of temperature. The best values of IE were gotten 81% at 298 K and 0.5 $\frac{g}{l}$ AO which means highest resistance to corrosion. The Langmuir isotherm was fitted to the adsorption of a banner.

Essential oil from fennel (Foeniculum Vulgare) (FM) was examined as a corrosion inhibitor of carbon steel in 1 M HCl by Hammuti et al. [150] using EIS, tafel polarization methods, and weight loss measurements. The analysis of FM oil by GC and GC/MS revealed that main compositions were limonene and β-pinene. Based on obtained results, FM functioned as a mixed-type banner by affecting both anodic and cathodic reactions. IE was achieved a maximum of 76% at 3 $\frac{ml}{l}$ but reduced with the rise of temperature.

Majidi et al. [151] examined inhibition effect of Mentha Spicata Essential oil on the corrosion of steel in 1 M HCl solution applying electrochemical polarization and weight loss measurements. Mentha Spicata acted as a mixed kind inhibitor and carvone was the major component of oil. IE enhanced with concentration of barrier to reach 97% at 2.00 gl^{-1} and decreased with rise of temperature. The adsorption of oil on the steel surface obeyed the Langmuir isotherm with physisorption mechanism.

Abdallah et al. [152] examined inhibition efficiency of natural black cumin oil on corrosion of nickel in 0.1 M HCl solution by means of galvanostatic and potentiodynamic anodic polarization techniques. It was detected that IE increased with the rise of inhibitor concentration. The adsorption of this compound obeyed Langmuir isotherm.

Thymus Satureioids (TS) oil was studied as green inhibitor for tinplate in 0.5 M HCl solution by Hammoutic et al. [153] using polarization curves. It was detected that TS acted as mixed inhibitor without changing the mechanism of hydrogen evolution reaction. Inhibition performance of the oil flourished with concentration to reach 87% at 6 gl^{-1} but diminished at high temperature to attain 75% at 65°C. It was found that TS functioned as a good inhibitor with a physical adsorption mechanism.

Hammouti et al. [154] investigated inhibition performance of natural oil extract from Pennyroyal Mint (Mentha pulegium, PM) on the corrosion of steel in 1 M HCl solution by means of weight loss, electrochemical polarization, and EIS. Based on obtained conclusions, natural oil worked as a cathodic inhibitor with modifying the hydrogen reduction mechanism. IE increased with oil content to reach its maximum 80% at 2.76 gl^{-1}. It was also enhanced by rise of temperature which indicated chemical adsorption. The adsorption of PM fitted well with Frumkin isotherm.

Summary of the information in this section appears in Table 6.1.

6.2 IMPRESSION OF NATURAL OILS IN H_2SO_4 MEDIA

The inhibition effect of essential oil of salvia aucheri mesatlantica was evaluated on the corrosion of steel in 0.5 M H_2SO_4 applying electrochemical polarization and weight loss methods by Majidi et al. [155]. It was found that inhibition efficiency of natural oil is related to its major component named camphor. IE increased by concentration of natural oils and temperature. The inhibitor acted as mixed-type, predominantly anodic effect. The adsorption of the barrier followed Langmuir isotherm.

Salghi et al. [156] tested corrosion behavior of C38 steel in 0.5 M H_2SO_4 by essential oil of Eucalyptus globulus (Myrtaceae) applying weight loss, potentiodynamic polarization, and EIS techniques. Eucalyptus globulus behaved as a mixed-type inhibitor. IE enhanced with the rise of concentration of Eucalyptus globulus but reduced with increase in temperature. The adsorp-

Table 6.1: Influence of natural oils in HCl solution (*Continues*)

NO	Metal	Inhibitor	Additives	Methods	Findings	Ref
91	Mild steel	Natural oil (Arachis hypogeae)		Gravimetric, potentiodynamic polarization, SEM-EDX	Mixed-type inhibitor, reduction of corrosion rate and increase of IE with addition of inhibitor, Langmuir isotherm	[3]
92	Mild steel	Avogadro natural oil		Gravimetric and potentiodynamic polarization	Inhibition efficiency increased by concentration of barrier and reached 81% at 405 $\frac{g}{V}$ Avogadro natural oil addition, Langmuir isotherm	[148]
93	Tinplate	Artemisia essential oil (AO)		Potentiodynamic polarization	Mixed kind inhibitor, The best values of IE, 81% at 298 k and 0.5 $\frac{g}{l}$ AO, Langmuir isotherm	[149]
94	Carbon steel	Essential oil from fennel (Foeniculum Vulgare) (FM)		Electrochemical impedance spectroscopy (EIS), tafel polarization methods, weight loss	Limonene and β-pinene, main compounds, mixed type inhibitor by affecting both anodic and cathodic reactions, maximum of IE: 76% at 3 mL/L, reduction of IE with rise of temperature	[150]
95	Steel	Mentha Spicata Essential oil		Electrochemical polarization, weight loss	Mixed kind inhibitor, carvon, the major component of oil, enhancement of IE with concentration of barrier to reach 97% at 2.00gl^{-1} and decrease with rise of temperature, Langmuir isotherm with physisorption mechanism	[151]
96	Nickel	Black cumin oil		Galvanostatic and potentiodynamic anodic polarization	Langmuir isotherm	[152]
97	Tinplate	Thymus Satureioids (TS) oil		Polarization curves	Mixed inhibitor without changing the mechanism of hydrogen evolution reaction, 87% IE at 6gl^{-1} inhibitor, physical adsorption mechanism	[153]

Table 6.1: (*Continued*) Influence of natural oils in HCl solution

| 98 | Steel | Natural oil extract from Pennyroyal Mint (Mentha pulegium, PM) | | Weight loss, electrochemical polarization, EIS | Cathodic inhibitor with modifying the hydrogen reduction mechanism, 80% IE at 2.76 gl^{-1} inhibitor, enhancement of IE by the rise of temperature, chemical adsorption, Frumkin isotherm | [154] |

tion of essential oils on a metal surface was spontaneous, physical, and followed the Langmuir isotherm.

Application of the essential oil Artemisia herba alba (Art) as green corrosion inhibitor for steel in 0.5 M H$_2$SO$_4$ was tested by Ouachikh et al. [157] with help of weight loss and electrochemical polarization methods. Collected results exhibited chrysanthenone and camphor were the main components which reduced corrosion rate by cathodic action. It means Art acted as a cathodic inhibitor with modifying hydrogen reduction mechanism. IE reached its maximum (74%) at 1 gl^{-1} and enhanced with both concentrations of Art and temperature.

Summary of the information in this section appears in Table 6.2.

6.3 EFFECT OF NATURAL OILS IN NACL SOLUTION

Roasted Elaeis guineensis was tested as a corrosion inhibitor on the corrosion process of Extruded AA6063 Al-Mg-Si alloy in simulated 3.5% NaCl using linear potentiodynamic polarization and gravimetric techniques by Fayomi and Popoola [158]. Elaeis guineensis essential natural oil was confirmed to be a good inhibitor for Al-Mg-Si, and functioned as a mixed-type banner. IE improved about 98% at an inhibitory concentration of 15%. The adsorption of inhibitor followed the Langmuir isotherm.

Corrosion inhibition of thermally pre-aged Aluminum-Silica-Magnesium (Al-Si-Mg) alloy with natural Avogadro oil in 3.5% NaCl solution was investigated by Abulwahab et al. [159] utilizing linear polarization technique. Avogadro oil behaved as a mixed inhibitor. IE and corrosion resistance of thermally pre-aged Al-Si-Mg increased with inhibitor concentration and reached 46.7%, 58%, and 71% by addition of 1.5, 3.0, 4.5$\frac{g}{v}$ Avogadro oil. The adsorption behavior of barrier followed the Langmuir isotherm.

Youssef et al. [160] found that cu 10Ni alloys are prone to stress corrosion cracking (SCC) in sulfide-polluted salt water (3.5% NaCl solution). The addition of glycine (Gly) as a potential environmentally friendly corrosion inhibitor to the examined solution led to the inhibition of SCC of alloy with raising the time to failure and changing the type of fracture from fragile

Table 6.2: Impression of natural oils in H_2SO_4 media

NO	Metal	Inhibitor	Additives	Methods	Findings	Ref
99	Steel	Essential oil of salvia aucheri mesatlantica		Electrochemical polarization, weight loss	Mixed-type predominantly anodic effect, camphor, major and active component of inhibitor, increase of IE by temperature, Langmuir isotherm,	[155]
100	C38 steel	Essential oil of Eucalyptus globulus (Myrtaceae)		Weight loss, potentiodynamic polarization, EIS	Mixed-type inhibitor, enhancement of IE with rise of concentration of Eucalyptus globulus but reduction with increase in temperature, spontaneous and physical adsorption following Langmuir isotherm	[156]
101	Steel	Essential oil of Artemisia herba alba (Art)		Weight loss, electrochemical polarization	Chrysanthenone and camphor, the main components, reduction of corrosion rate by cathodic action, the maximum of IE: 74% at 1 gl-1 and enhancement of that with both concentration of Art and temperature	[157]

transgranular cracking to ductile failure. It was verified by applying electrochemical measurements, including potentiodynamic polarization and FEM. Gly was discovered to function as a mixed-type inhibitor. By addition of KI to the solution, IE enhanced from 73% to 87% because synergistic effect between gly and I^- ions resulted to fixed adsorption of gly molecules on the metal surface.

Halambek et al. [161] examined the influence of Lavandula angustifolia L. oil on corrosion of Al-3Mg alloy in 3% NaCl solution by means of weight loss, polarization measurements, and SEM. L. angustifolia L. oil performed as a mixed banner. The adsorption of barrier was both physical and chemical as well as obeying the Langmuir isotherm.

Porcayo-Calderon et al. [162] evaluated a hydroxyethyl-imidazoline derivated based on coffee oil as corrosion inhibitor for carbon steel in CO_2-saturated (3% Nacl+10% diesel) emulsion at 50°C by using the EIS method. Results showed that this inhibitor decreased corrosion rate by over 99.9%. Optimal concentration was 10 ppm. Higher or lower concentration enhanced the corrosion rate due to decreasing the surface area covered by inhibitor and formation of unsupported sites on the metal. In fact, added concentration was not sufficient to form a protective layer or existence of an extra banner favors the appearance of electrostatic repulsion forces between the negative charges causing a desorption of the banner molecules leading to the formation of unsupported sites on the metal.

Summary of these investigations is shown in Table 6.3.

6.4 EFFECT OF NATURAL OIL IN OTHER CORROSIVE MEDIA

Abdallah et al. [73] investigated natural oils including parsley, lettuce, sesame, arugula, and sweet almond oils as inhibitors on 304 stainless steel in 0.1 M NaOH solution. Researchers were used galvanostatic, potentiodynamic anodic polarization, and EIS. The outcomes showed IE depends on the nature of oil and its concentration. In fact, by increasing the oil concentration, a decrease was observed in the corrosion current density, an increase in IE, an increase in the surface coverage, a movement in the pitting potential toward more positive values, and an increase in the charge transfer resistance. The inhibition efficiency of inhibitors was in the order: sesame oil > lettuce oil > sweet almond oil > parsley oil > arugula oil. The adsorption of inhibitors obeyed Langmuir isotherm. With increasing the concentration of inhibitor, charge transfer resistance increased while the capacitance of double layer reduced.

Some natural oils including sesame oil, watercress oil, wheat germ oil, and almond oil, natural inhibitors for the corrosion of nickle electrode in 1×10^{-2} M NaOH solution, was studied by Abdallah et al. [163]. Circuit potential measurements, galvanostatic, and potentiostatic polarization methods were utilized. Inhibition efficiency increased with concentration of these oils and followed in the order: sesame oil > watercress oil > wheat germ oil > almond oil. Adsorption behavior of inhibitors obeyed Freundlish isotherm.

Table 6.3: Effect of natural oils in NaCl solution

NO	Metal	Inhibitor	Additives	Methods	Findings	Ref
102	AA6063 Al-Mg-Si alloy	Roasted Elaeis guineensis		Linear potentiodynamic polarization, gravimetric	Mixed type banner, 98% IE at 15% inhibitor concentration, Langmuir isotherm	[158]
103	Aluminum-Silica-Magnesium (Al-Si-Mg) alloy	Avogadro oil		Linear polarization	Mixed inhibitor, The values of IE and corrosion resistance: 46.7%, 58%, and 71% by addition of 1.5, 3.0, 4.5 $\frac{g}{v}$ Avogadro oil, Langmuir isotherm	[159]
104	cu 10Ni alloys	Glycine (gly)	KI	Electrochemical measurements including potentiodynamic polarization and FEM	Mixed type inhibitor, increase of IE with addition of KI to the solution from 73–87% because of synergistic effect between gly and I- ions,	[160]
105	Al-3Mg alloy	Lavandula angustifolia L. oil		Weight loss, polarization measurements, SEM	Mixed banner, both physical and chemical adsorption, Langmuir isotherm	[161]
106	Carbon steel	Hydroxyethyl-imidazoline derivated based on coffee oil		EIS	10 ppm optimal concentration of inhibitor, increasing corrosion rate at higher or lower concentration due to decreasing the surface area covered by inhibitor and formation of unsupported sites on the metal	[162]

Galvanostatic and potentiodynamic anode polarization methods were employed by Abdallah et al. [164] to study inhibition efficiency of some natural oils containing parsley, lettuce, and radish oils on corrosion of carbon steel L-52 in 0.5 M NaOH. Inhibitors played as mixed type. IE increased by rise of inhibitors concentration following lettuce oil > parsley oil > radish oil. The Langmuir isotherm was fitted to the adsorption of barriers.

Electrochemical techniques, EIS, and polarization were carried out to study and compare the corrosive power of copper in 2 M HNO_3 with essential oil of Thymus Satureoides and its main component borneol by Houbairi et al. [165]. Results showed Thymus Satureoides acted as an effective banner. The inhibition action was related to different chemical materials and main product (borneol). IE increased by rising in banner concentration to get 89.02% at 1,200 ppm for the essential oil of thymus satureoides and 69.72% at 1,600 ppm for the main ingredient. Tested inhibitors acted as a mixed type with a significant cathodic efficiency for the essential oil. Addition of inhibitor increased the corrosion resistance of copper. Inhibitors functioned by reduction of active area without changing the mechanism of anodic and cathodic processes.

Houbairi et al. [166] worked on corrosion behavior of copper in nitric acid media (2M) by essential oil of Thyme Morocco. Weight loss and polarization techniques were used. It was found that the inhibition property of Thymus Satureoides oil was predominantly thanks to a synergy between chemical materials and not just to its main compound (borneol). IE was achieved 89.02% at 1,200 ppm for essential oil of Thymus Satureoides and 69.72% at 1,600 ppm for major substance alone. Mentioned inhibitor played as mixed kind banner with main cathodic efficiency for essential oil. IE decrease by rising temperature. Two inhibitors adsorbed physically according to Langmuir isotherm.

Hammouti et al. [167] studied the inhibition effect of Artemisia oil (Ar) on corrosion of steel in 2 M H_3PO_4 using weight loss, electrochemical polarization, and EIS methods. It was found that Ar worked as a cathodic inhibitor with modifying the hydrogen reduction mechanism. IE increased with inhibitor concentration to reach a maximum value 79% at 6 gl^{-1} and reduced with a rise in temperature.

Evaluation of rosemary oil as a green corrosion inhibitor for steel in 2 M H_3PO_4 solution was done by Bendahou et al. [168] using gravimetric, polarization, and EIS methods. It was found that rosemary oil acted as a cathodic inhibitor modifying the hydrogen reduction mechanism. IE increased with concentration to reach 73% at 10 gl^{-1} of inhibitor but reduced with the rise in temperature.

Summary of the information in this section appears in Table 6.4.

Considerable efforts are deployed to find suitable compound to be used as corrosion inhibitor in various corrosive media to stop or delay to the maximum attack of a metal. Nevertheless, the known hazardous effects of the most synthetic inhibitors and the need to develop environmentally friendly processes, researchers are focused on the use of natural products.

Table 6.4: Effect of natural oil in other corrosive media (*Continues*)

NO	Metal	Medium	Inhibitor	Additives	Methods	Findings	Ref
107	304 stainless steel	0.1 M NaOH	Natural oils including parsley, lettuce, sesame, arugula, sweet almond oils		Galvanostatic, potentiodynamic anodic polarization, EIS	The inhibition efficiency of inhibitors: sesame oil > lettuce oil > sweet almond oil > parsley oil > arugula oil, Langmuir isotherm	[73]
108	Nickle	1×10^{-2} M NaOH	Sesame oil, water cress oil, wheat germ oil, almond oil		Circuit potential, galvanostatic, and potentiostatic polarization	Inhibition efficiency of oils: sesame oil > water cress oil > wheat germ oil > almond oil, Freundlish isotherm	[163]
109	Carbon steel L-52	0.5 M NaOH	Parsley, lettuce, and radish oils		Galvanostatic and potentiodynamic anode polarization	Mixed-type, inhibition efficiency of inhibitors: lettuce oil > parsley oil > radish oil, Langmuir isotherm	[164]
110	Copper	2 M HNO$_3$	Essential oil of Thymus Satureoides and its main component borneol		Electrochemical techniques, EIS, polarization	Mixed-type inhibitors with major cathodic effect for essential oil, maximum of IE:89.02% and 69.72% for essential oil and its main ingredient respectively	[165]
111	Copper	2 M HNO$_3$	Essential oil of Thyme Morocco		Weight loss, polarization	Mixed kind banner with main cathodic efficiency for essential oil, main active compound of banner (borneol), physical adsorption according to Langmuir isotherm	[166]

Table 6.4: (*Continued*) Effect of natural oil in other corrosive media

112	Steel	2 M H_3PO_4	Artemisia oil (Ar)		Weight loss, electrochemical polarization, EIS	Cathodic inhibitor with modifying the hydrogen reduction mechanism, increase of IE with inhibitor concentration to reach a maximum value 79% at 6 gl^{-1} and reduction of that with rise of temperature	[167]
113	Steel	2 M H_3PO_4	Rosemary oil		Gravimetric, polarization, EIS	Cathodic inhibitor modifying the hydrogen reduction mechanism, 73% IE at 10 gl^{-1} of inhibitor, reduction of IE with rise of temperature	[168]

Natural plants in the form of extract, oils, or pure compounds may play major roles in keeping the environment healthier, safe, and under pollution control. Encouraging results obtained by natural oils as corrosion inhibitors permit the testing of more substance oils.

CHAPTER 7

A Journey to the Natural Corrosion Inhibitors in Corrosive Environments

7.1 EFFECT OF NATURAL INHIBITORS IN HCL MEDIA

Iota-carrageenan, a natural polymer, was studied as a corrosion inhibitor on Al in the presence of pefloxacin mesylate as zwitterion mediator in 2 M HCl solution. Fares et al. [54] found out a remarkable enhancement in the inhibition performance values from 66.7% in the absence of mediator to 91.8% in the presence of mediator. The formation of coherent physical adsorption layer on the surface of Al led to high efficiency. Adsorption isotherms in the absence and presence of pefloxacin were fitted in Langmuir. The exposure of Al surface to 2.0 M HCl for 2 h in the presence of 100 ppm iota-carrageenan at 400 ppm pefloxacin, turned out with smaller but stable regular shape broken layers.

Thermodynamic and electrochemical investigations of (9-[(R) 2 [[bis [[(isopropoxycarbonyl) oxy] methoxy] phosphinyl] methoxy] propyl] adenine fumarate)] (Tenvir) as a green corrosion inhibitor for mild steel in 1 M HCl was carried out by Quraishi et al. [169]. Weight loss, potentiodynamic polarization, and EIS methods were applied. Tenvir functioned as an effective and mixed inhibitor because it quenched both anodic and cathodic reactions of mild steel corrosion in 1 M HCl. Inhibition efficiency of Tenvir enhanced by an increase in concentration and reached to its highest (95.7%) at 400 ppm. The adsorption of Tenvir followed the Langmuir isotherm. According to negative values of ΔG°_{ads}, the adsorption process is spontaneous.

Chitosan (a naturally occurring polymer) in unmodified form was evaluated to reduce corrosion attack of mild steel in HCl solution by Umoren et al. [170] by means of gravimetric, potentiodynamic polarization, EIS, SEM, and UV-visible analysis. Chitosan acted as effective inhibitor. IE increased with a rise in temperature up to 96% at 60°C and then reduced 93% at 70°C while it rarely enhanced by rising in chitosan concentration. It was discovered that the mentioned banner played as a mixed inhibitor. Chemical adsorption of inhibitor constituents and the Langmuir isotherm was suggested for the adsorption of chitosan to mild steel surface.

The inhibitory action of new drug, namely Abacavir Sulfate ((4-(2- amino-6-(cyclopropyl amino)-9H-purin-9-yl) cyclopent-2-enyl) methanol sulfate) on mild steel in 1 M HCl was explored by Quraishi et al. [171] using potentiodynamic polarization, linear polarization, EIS, and weight loss measurements. According to the achieved results, the Abacavir performed as a mixed-type banner. IE enhanced by concentration of banner and maximum efficiency (98%) was observed at 400 ppm concentration. The adsorption of Abacavir on mild steel surface followed the Langmuir isotherm. The negative values of ΔG°_{ads} showed spontaneous process of adsorption of inhibitor on mild steel surface in 2 M HCl.

Quinazoline derivatives as green corrosion inhibitors for carbon steel in 2 M HCl were studied by Fouda et al. [172]. Potentiodynamic polarization, EIS, SEM, and EDS methods were exerted to explore inhibition efficiency of (4-(4-amino-6,7-dimethoxy quinazolin-2-yl)piperazin-1-yl) (2,3-dihydrobenzo [b] [1,4] dioxin-2-yl) methanone, (4-(4-amino-6,7-dimethoxy quinazolin-2-yl) piperazin-1-yl) (tetra hydro furan-2-yl) methanone, (4-(4-amino-6,7-dimethoxy quinazolin-2-yl) piperazin-1-yl) (furan-2-yl) methanone. All of the derivatives functioned very well and better than the passive oxide film. Based on polarization curves quinazoline derivatives played as mixed kind inhibitors, but cathode is more polarized than anode. IE increased with an increase in concentration while reduced by a rise in temperature. Inhibition effect of tested compounds was in order: 1 > 2 > 3. The adsorption of mentioned banners obeyed the Temkin isotherm.

Grape pomace extract was proved to have inhibition efficiency for carbon steel in 1 M HCl by da Rocha et al. [173] applying weight loss, potentiodynamic polarization curves, EIS measurements, and surface analysis including SEM. IE increased with concentration and was inversely with temperature. Flavonoids were responsible for the inhibiting action of grape pomace extract. The adsorption banner followed the Langmuir isotherm. SEM studies showed the resistance of a soft surface on C-steel after adding grape pomace was likely because of the formation of adsorptive layer of phenolic materials with electrostatic character.

Menthol has been detected to be an effective volatile corrosion inhibitor for copper in 0.0 1 N HCl. Premkumar et al. [174] investigated the inhibition effect of Menthol using gravimetric and electrochemical methods (potentiodynamic polarization and EIS). Based on the obtained results, the maximum inhibition efficiency was attained at concentration $1,000 \frac{mg}{sq.ft}$. The Temkin isotherm was described for the adsorption of the mentioned barrier on copper surface.

Effect of pulegone and pulegone oxide on corrosion of steel in 1 M HCl was examined by Faska et al. [175] using weight loss, potentiodynamic polarization, and EIS. Results showed IE enhanced with inhibitor content to reach 81 and 75% at 5 g dm^{-3} for pulegone and pulegone oxide. The growth of temperature caused an enhancement in inhibition performance of mentioned barriers.

Inhibition of the corrosion of steel in 1 M HCl by eugenol (Eug) and its derivative acetyleugenol (Ac Eug) was explored by Hammouti et al. [176] with the help of weight loss, electrochemical polarization, and EIS. It was found inhibition performance of Eug and Ac Eug grew with inhibitors concentration to reach a maximum value 91% at 0.173 gl^{-1} Ac Eug. IE increased with a rise in temperature. Eugenol and its derivative functioned as mixed inhibitors without changing the hydrogen reduction mechanism. The Langmuir isotherm was described for the adsorption of inhibitors on the steel.

Quinine, a natural product, was studied as a corrosion inhibitor for low carbon steel in 1 M HCl solution by Awad [177] using potentiodynamic polarization and EIS. IE increased with inhibitor concentration and attained a near constant value in the concentration range 0.48 mM and above. Corrosion of steel was monitored by a charge transfer process. The adsorption of inhibitor was physically and followed the Langmuir isotherm.

Summary of the information in this section appears in Table 7.1.

7.2   INFLUENCE OF NATURAL BANNERS IN HCL AND H₂SO₄ SOLUTION

Cefazolin (CZ) and cefotaxime (CT) were proved to be good and environmently corrosion banners on carbon steel in 0.5 M H₂SO₄ solution as shown by Nazeer et al. [178]. It was discovered CZ and CT acted as mixed inhibitors because both anodic and cathodic reactions were prolonged. The inhibition was found to follow the Langmuir isotherm and was performed by chemisorptions and physisorption. The adsorption of banners was spontaneous because of negative values of $\Delta G°_{ads}$. Maximum inhibition efficiency was achieved 99.6% (5×10^{-4} M CZ) and 90.9% (7×10^{-4} M CT). IE decreased with an increase of temperature.

Popoola et al. [179] tested corrosion inhibition effect of sesamum indicum on mild steel in 2 M HCl/H₂SO₄ solution by means of gravimetric and linear polarization methods. It was found that IE and corrosion resistance enhanced with inhibitor concentration in both mentioned environments. Sesamum indicum functioned as a mixed type. The adsorption of barrier obeyed the Langmuir isotherm.

Lornoxicam and Tenoxicam Drugs as a green corrosion inhibitor for carbon steel in 0.5 M HCl and 1 M H₂SO₄ solution were used by the Fouda et al. [180], applying weight loss, potentiodynamic polarization, and EIS. It was found that inhibitors acted as mixed types. IE enhanced with a rise of barrier concentration but reduced with temperature. The adsorption of Lornoxicam and Tenoxicam were physical and chemical, and obeyed the Temkin isotherm.

Summary of the information in this section appears in Table 7.2.

7.3   IMPRESSION OF NATURAL BANNERS IN OTHER CORROSIVE ENVIRONMENTS

Ibrahim et al. [181] studied inhibition efficiencies of kraft lignin (KL) and soda lignin (SL) on the corrosion of mild steel in 3.5% $\frac{w}{v}$ NaCl at pH 6 and 8. KL and SL were detected to be mixed inhibitors. The maximum inhibition efficiency of SL and KL was 89% and 87% at pH 6 as well as 92% and 90% at pH 8, respectively. SL showed better IE in comparison to KL at both pH levels and lower performance at a higher pH. The Langmuir isotherm was described for the adsorption of KL and SL. All of these results were obtained by employing weight loss,

Table 7.1: Effect of natural inhibitors in HCl media (*Continues*)

NO	Metal	Inhibitor	Additives	Methods	Findings	Ref
114	Aluminum (Al)	Iota-carrageenan	Pefloxacin mesylate as zwitterion mediator	Weight loss, thermodynamic parameters	Remarkable enhancement in the inhibition performance values from 66.7% in the absence of mediator to 91.8% in the presence of mediator, physical adsorption, Langmuir isotherm	[54]
115	Mild steel	(9-[(R) 2 [[bis [[(isopropoxycar-bonyl) oxy] me-thoxy] phosphinyl] methoxy] propyl] adenine fumarate)] (Tenvir)		Weight loss, potentiodynamic polarization, EIS, SEM-EDX	Mixed inhibitor, highest IE at 400 ppm: 95.7%, Langmuir isotherm, spontaneous adsorption	[169]
116	Mild steel	Chitosan (a naturally occurring polymer)		Gravimetric, potentiodynamic polarization, EIS, SEM, UV-visible	Mixed inhibitor, increase of IE with a rise in temperature up to 96% at 60°C and then reduction of that 93% at 70°C, chemical adsorption, Langmuir isotherm	[170]
117	Mild steel	Abacavir Sulfate ((4-(2- amino-6-(cyclopropyl amino)-9H-purin-9-yl) cyclopent-2-enyl) methanol sulfate)		Potentiodynamic polarization, linear polarization, EIS, and weight loss	Mixed-type inhibitor, the maximum efficiency (98%) at 400 ppm concentration, Langmuir isotherm, spontaneous adsorption	[171]
118	Carbon steel	Quinazoline derivatives		Potentiodynamic polarization, EIS, SEM, and EDS	Mixed kind inhibitors, but cathod is more polarized than anode, increase of IE with an increase in concentration and reduction of that by rising in temperature, Temkin isotherm	[172]

Table 7.1: (*Continued*) Effect of natural inhibitors in HCl media

119	Carbon steel	Grape pomace		Weight loss, potentiodynamic polarization curves, EIS, SEM	Flavonoids, the main active element and responsible for inhibition action, Langmuir isotherm, increase of IE with concentration and reduction of that with temperature	[173]
120	Copper	Menthol		Gravimetric and electrochemical methods (potentiodynamic polarization and EIS)	Temkin isotherm	[174]
121	Steel	Pulegone and pulegone oxide		Weight loss, potentiodynamic polarization, EIS	Increase of IE with inhibitor content to reach 81 and 75% at 5 g dm^{-3} for pulegone and pulegone oxide, increase of IE with rising temperature	[175]
122	Steel	Eugenol (Eug) and its derivative acetyleugenol (Ac Eug)		Weight loss, electrochemical polarization, EIS	Mixed inhibitors without changing the hydrogen reduction mechanism, increase of IE with rise of temperature, Langmuir isotherm	[176]
123	Low carbon steel	Quinine		Potentiodynamic polarization, EIS	Physical adsorption, Langmuir isotherm	[177]

Table 7.2: Influence of natural banners in HCl and H_2SO_4 solutions

NO	Metal	Inhibitor	Additives	Methods	Findings	Ref
124	Carbon steel	Cefazolin (CZ) and cefotaxime (CT)		Potentiodynamic polarization, EIS, EFM	Mixed inhibitors, maximum inhibition efficiency 99.6% (5×10^{-4} M CZ) and 90.9% (7×10^{-4} M CT), decrease of IE with increasing temperature, spontaneous, chemical and physical adsorption, Langmuir isotherm	[178]
125	Mild steel	Sesamum indicum		Gravimetric and linear polarization	Mixed type, Langmuir isotherm	[179]
126	Carbon steel	Lornoxicam and Tenoxicam Drugs		Weight loss, potentiodynamic polarization, EIS	Mixed types, enhancement of IE with rise of barrier concentration but reduction of that with temperature, physical and chemical adsorption, Temkin isotherm	[180]

electrochemical techniques consisting of potentiodynamic polarization and EIS, and surface analysis.

Citric acid as a natural corrosion inhibitor was applied by Kardas et al. [182] in order to protect aluminium against corrosion in 2 M NaCl solution (pH2) utilizing potentiodynamic polarization, linear polarization resistance (LPR), and EIS. Based on gathered conclusions, citric acid functioned as a mixed inhibitor with mainly cathodic action. IE increased with the concentration of inhibitor up to 1×10^{-5} M which made the best protection but decreased at higher concentration.

Vrsalovic et al. [183] tested salvia officinalis L. honey as corrosion inhibitor for cu Ni Fe alloy in 0.5 M NaCl solution by means of potentiodynamic polarization and EIS. According to the results, the honey functioned as mixed-type inhibitor. IE increased with a rise of inhibitor concentration but diminished with a rise of temperature of the electrolyte. The maximum value of IE was nearly achieved 70% at 1,200 ppm of honey. The adsorption of organic compound from honey on the cu Ni Fe surface followed Langmuir isotherm.

Zaafarany [184] proved Alginate (Alg) and Pectates (pec) water-soluble natural polymer anionic polyelectrolytes are impressive inhibitors for corrosion of Al in NaOH solution and diminished its corrosion rate. Gasometric and weight loss methods were done to perform this research. Inhibition efficiency of pect was more than Alg which maybe it is because of geometrical configuration of the functional groups of banners that have main role in the value of IE.

Researchers supposed natural or synthetic polymers including the same functional groups have the same corrosion action and mechanism.

Chrzescijanska and Kusmierek [185] explored tannic acid as natural corrosion inhibitor for two series of tested samples (metals and alloys) utilizing liner polarization resistance. Electrochemical behavior of metal samples in tannic acid solutions indicated no notable changes in E_{corr} but obvious changes in corrosion rate. Corrosion rate depends on the concentration of tannic acid. Inhibition efficiency of tannic acid toward metals by regarding its different concentration and immersion numbers was in order: $Cu > Zn > Fe$. Electrochemical behavior of alloy samples in tannic acid solutions was shown a clear decrease in E_{corr} values in the item of Inconel alloy and brass in comparison with ferritic steel samples and explicit variations in i_{corr} as well as R_p values were observed in the case of all studied alloys (steel, Inconel, brass). According to inhibition efficiency values of tannic acid for alloys, this compound can act as corrosion inhibitor. Inhibition efficiency of tannic acid solutions with different concentration toward all tested alloys was followed: Inconel > steel > brass. More inhibition efficiency of Inconel is due to higher content of Cr and lower content of Fe in comparison with steel, and the least protected alloy was brass thanks to the content of Zn which suffers from corrosion attack, although in higher concentration of tannic acid solutions (above $20\frac{mg}{dm^3}$) brass is more supported against corrosion than other alloys. By considering the number of immersion in acid solution IE was in order: brass > Inconel > steel.

Summary of the information in this section appears in Table 7.3.

To summarize, corrosion is the deterioration of materials by chemical interaction with their environment. Generally, the term corrosion refers to metals. The protection of metals and alloys against corrosion can be achieved either by special treatment of the medium to depress its aggressiveness or by introducing into it small amounts of special substances called corrosion inhibitors.

A number of organic and inorganic compounds are known to be applicable as corrosion inhibitors in corrosive environments. Unfortunately, these compounds are very expensive. So, development of novel corrosion inhibitors of natural source and none-toxic type, which don't have heavy metals, has been considered more important and desirable. The natural inhibitors are green inhibitors because of their environmentally friendly nature. Green inhibitors have a promising future for the quality of the environment. In addition, they are biodegradable and a renewable source of materials. Regarding the importance of using natural banners, based on the mentioned reasons, it is essential to pay attention and apply them more in future investigations.

Table 7.3: Impression of natural banners in other corrosive environment

NO	Metal	Medium	Inhibitor	Additives	Methods	Findings	Ref
127	Mild steel	$3.5\%\frac{w}{v}$ NaCl	Kraft lignin (KL) and soda lignin (SL)		Weight loss, electrochemical techniques consisting of potentiodynamic polarization and EIS, also surface analysis pH 6, 8	Mixed inhibitors, maximum inhibition efficiency of SL and KL: 89 and 87% at pH 6 as well as 92 and 90% at pH 8, respectively, better IE for SL in comparison with KL, Langmuir isotherm	[181]
128	Aluminium	2 M NaCl	Citric acid		Potentiodynamic polarization, linear polarization resistance (LPR), EIS	Mixed inhibitor with mainly cathodic action	[182]
129	Cu Ni Fe alloy	0.5 M NaCl	Salvia officinalis L. honey		Potentiodynamic polarization, EIS	Mixed-type inhibitor, reduction of IE with rise of temperature, the maximum value of IE: 70% at 1200 ppm of honey, Langmuir isotherm	[183]
130	Aluminum (Al)	NaOH	Alginate (Alg) and Pectates (pec)		Gasometric, weight loss	More inhibition efficiency of pect than Alg	[184]
131	Two series of tested samples (metals and alloys)	Tannic acid solution	Tannic acid		Liner polarization resistance	IE of tannic acid towards metals by regarding its different concentration and immersion numbers: Cu > Zn > Fe, IE of tannic acid solutions with different concentration towards all tested alloys: Inconel > steel > brass, IE of alloys by considering the number of immersion in acid solution: brass > Inconel > steel	[185]

References

[1] N. Soltani, et al., Electrochemical and quantum chemical calculations of two schiff bases as inhibitor for mild steel corrosion in hydrochloric acid solution. *Iranian Journal of Analytical Chemistry*, 2(1), pp. 22–35, 2015. 1, 6

[2] M. Behpour, et al., The effect of two oleo-gum resin exudate from Ferula assa-foetida and Dorema ammoniacum on mild steel corrosion in acidic media. *Corrosion Science*, 53(8), pp. 2489–2501, 2011. DOI: 10.1016/j.corsci.2011.04.005. 1, 9, 18

[3] O. Fayomi, M. Abdulwahab, and A. Popoola, Electro-oxidation behaviour and passivation potential of natural oil as corrosion inhibitor in hydrochloric acid environment. *International Journal of Electrochemical Science*, 8, pp. 12088–12096, 2013. 1, 47, 49

[4] N. Soltani and M. Khayatkashani, Gundelia tournefortii as a green corrosion inhibitor for mild steel in HCl and H2SO4 solutions. *International Journal of Electrochemical Science*, 10(1), pp. 46–62, 2015. 1, 4, 6, 25, 30

[5] J. C. da Rocha, J. A. D. C. P. Gomes, and E. D'Elia, Corrosion inhibition of carbon steel in hydrochloric acid solution by fruit peel aqueous extracts. *Corrosion Science*, 52(7), pp. 2341–2348, 2010. DOI: 10.1016/j.corsci.2010.03.033. 1, 2, 3, 41, 42

[6] N. Soltani, et al., Silybum marianum extract as a natural source inhibitor for 304 stainless steel corrosion in 1.0 M HCl. *Journal of Industrial and Engineering Chemistry*, 20(5), pp. 3217–3227, 2014. DOI: 10.1016/j.jiec.2013.12.002. 1, 6, 9, 18

[7] A. Ostovari, et al., Corrosion inhibition of mild steel in 1 M HCl solution by henna extract: A comparative study of the inhibition by henna and its constituents (Lawsone, Gallic acid, α-d-Glucose and Tannic acid). *Corrosion Science*, 51(9), pp. 1935–1949, 2009. DOI: 10.1016/j.corsci.2009.05.024. 1, 2, 9, 18

[8] M. Desai, et al., Inhibition of corrosion of aluminium-51S in hydrochloric acid solutions. *Corrosion Science*, 16(1), pp. 9–24, 1976. DOI: 10.1016/s0010-938x(76)80003-0. 1

[9] H. El-Dahan, T. Soror, and R. El-Sherif, Studies on the inhibition of aluminum dissolution by hexamine–halide blends: Part I. Weight loss, open circuit potential and polarization measurements. *Materials Chemistry and Physics*, 89(2–3), pp. 260-267, 2005. DOI: 10.1016/j.matchemphys.2004.07.025. 1

[10] I. Obot, N. Obi-Egbedi, and S. Umoren, The synergistic inhibitive effect and some quantum chemical parameters of 2, 3-diaminonaphthalene and iodide ions on the hydrochloric acid corrosion of aluminium. *Corrosion Science*, 51(2), pp. 276–282, 2009. DOI: 10.1016/j.corsci.2008.11.013. 1

[11] A. El-Etre, Inhibition of acid corrosion of aluminum using vanillin. *Corrosion Science*, 43(6), pp. 1031–1039, 2001. DOI: 10.1016/s0010-938x(00)00127-x. 1

[12] S. Hassan, et al., Aromatic acid derivatives as corrosion inhibitors for aluminium in acidic and alkaline solutions. *Anti-Corrosion Methods and Materials*, 37(2), pp. 8–11, 1990. DOI: 10.1108/eb007261. 1

[13] I. Issa, M. Moussa, and M. Ghandour, A study on the effect of some carbonyl compounds on the corrosion of aluminium in hydrochloric acid solution. *Corrosion Science*, 13(10), pp. 791–797, 1973. DOI: 10.1016/s0010-938x(73)80016-2. 1

[14] M. Lashgari and A. M. Malek, Fundamental studies of aluminum corrosion in acidic and basic environments: Theoretical predictions and experimental observations. *Electrochimica Acta*, 55(18), pp. 5253–5257, 2010. DOI: 10.1016/j.electacta.2010.04.054. 1

[15] A. Fouda, et al., The role of some thiosemicarbazide derivatives in the corrosion inhibition of aluminium in hydrochloric acid. *Corrosion Science*, 26(9), pp. 719–726, 1986. DOI: 10.1016/0010-938x(86)90035-1. 1

[16] M. Moussa, et al., The effect of some hydrazine derivatives on the corrosion of Al in HCl solution. *Corrosion Science*, 16(6), pp. 379–385, 1976. DOI: 10.1016/0010-938x(76)90124-4. 1

[17] S. M. Hassan, et al., Studies on the inhibition of aluminium dissolution by some hydrazine derivatives. *Corrosion Science*, 19(12), pp. 951–959, 1979. DOI: 10.1016/s0010-938x(79)80086-4. 1

[18] G. Bereket and A. Yurt, The inhibition effect of amino acids and hydroxy carboxylic acids on pitting corrosion of aluminum alloy 7075. *Corrosion Science*, 43(6), pp. 1179–1195, 2001. DOI: 10.1016/s0010-938x(00)00135-9. 1

[19] M. Abdallah, Antibacterial drugs as corrosion inhibitors for corrosion of aluminium in hydrochloric solution. *Corrosion Science*, 46(8), pp. 1981–1996, 2004. DOI: 10.1016/j.corsci.2003.09.031. 1

[20] G. K. Gomma and M. H. Wahdan, Schiff bases as corrosion inhibitors for aluminium in hydrochloric acid solution. *Materials Chemistry and Physics*, 39(3), pp. 209–213, 1995. DOI: 10.1016/0254-0584(94)01436-k. 1

[21] A. Yurt, S. Ulutas, and H. Dal, Electrochemical and theoretical investigation on the corrosion of aluminium in acidic solution containing some Schiff bases. *Applied Surface Science*, 253(2), pp. 919–925, 2006. DOI: 10.1016/j.apsusc.2006.01.026.

[22] H. Ashassi-Sorkhabi, et al., The effect of some Schiff bases on the corrosion of aluminum in hydrochloric acid solution. *Applied Surface Science*, 252(12), pp. 4039–4047, 2006. DOI: 10.1016/j.apsusc.2005.02.148. 1

[23] Q. Zhang and Y. Hua, Corrosion inhibition of aluminum in hydrochloric acid solution by alkylimidazolium ionic liquids. *Materials Chemistry and Physics*, 119(1–2), pp. 57-64, 2010. DOI: 10.1016/j.matchemphys.2009.07.035. 1

[24] E. Oguzie, et al., Evaluation of the inhibitory effect of methylene blue dye on the corrosion of aluminium in hydrochloric acid. *Materials Chemistry and Physics*, 87(2–3), pp. 394-401, 2004. DOI: 10.1016/j.matchemphys.2004.06.003. 1

[25] T. Zhao and G. Mu, The adsorption and corrosion inhibition of anion surfactants on aluminium surface in hydrochloric acid. *Corrosion Science*, 41(10), pp. 1937–1944, 1999. DOI: 10.1016/s0010-938x(99)00029-3. 1

[26] S. S. A. El Rehim, H. H. Hassan, and M. A. Amin, The corrosion inhibition study of sodium dodecyl benzene sulphonate to aluminium and its alloys in 1.0 M HCl solution. *Materials Chemistry and Physics*, 78(2), pp. 337–348, 2003. DOI: 10.1016/s0254-0584(01)00602-2.

[27] A. Maayta and N. Al-Rawashdeh, Inhibition of acidic corrosion of pure aluminum by some organic compounds. *Corrosion Science*, 46(5), pp. 1129–1140, 2004. DOI: 10.1016/j.corsci.2003.09.009. 1

[28] S. S. A. El Rehim, H. H. Hassan, and M. A. Amin, Corrosion inhibition of aluminum by 1, 1 (lauryl amido) propyl ammonium chloride in HCl solution. *Materials Chemistry and Physics*, 70(1), pp. 64–72, 2001. DOI: 10.1016/s0254-0584(00)00468-5. 1

[29] X. Li, S. Deng, and H. Fu, Inhibition by tetradecylpyridinium bromide of the corrosion of aluminium in hydrochloric acid solution. *Corrosion Science*, 53(4), pp. 1529–1536, 2011. DOI: 10.1016/j.corsci.2011.01.032. 1

[30] E. F. El-Sherbini, S. Abd-El-Wahab, and M. Deyab, Studies on corrosion inhibition of aluminum in 1.0 M HCl and 1.0 M H2SO4 solutions by ethoxylated fatty acids. *Materials Chemistry and Physics*, 82(3), pp. 631–637, 2003. DOI: 10.1016/s0254-0584(03)00336-5. 1

[31] Y. Xiao-Ci, et al., Quantum chemical study of the inhibition properties of pyridine and its derivatives at an aluminum surface. *Corrosion Science*, 42(4), pp. 645–653, 2000. DOI: 10.1016/s0010-938x(99)00091-8. 1

[32] G. Bereket and A. Pinarbaşi, Electrochemical thermodynamic and kinetic studies of the behaviour of aluminium in hydrochloric acid containing various benzotriazole derivatives. *Corrosion Engineering, Science and Technology*, 39(4), pp. 308–312, 2004. DOI: 10.1179/174327804x13136. 1

[33] E. Khamis and M. Atea, Inhibition of acidic corrosion of aluminum by triazoline derivatives. *Corrosion*, 50(2), pp. 106–112, 1994. DOI: 10.5006/1.3293498. 1

[34] K. Khaled and M. Al-Qahtani, The inhibitive effect of some tetrazole derivatives towards Al corrosion in acid solution: Chemical, electrochemical and theoretical studies. *Materials Chemistry and Physics*, 113(1), pp. 150–158, 2009. DOI: 10.1016/j.matchemphys.2008.07.060. 1

[35] S. Deng and X. Li, Inhibition by Jasminum nudiflorum Lindl. leaves extract of the corrosion of aluminium in HCl solution. *Corrosion Science*, 64, pp. 253–262, 2012. DOI: 10.1016/j.corsci.2012.07.017. 1, 4, 10, 18

[36] G. Ilevbare and G. Burstein, The inhibition of pitting corrosion of stainless steels by chromate and molybdate ions. *Corrosion Science*, 45(7), pp. 1545–1569, 2003. DOI: 10.1016/s0010-938x(02)00229-9. 1

[37] W. Badawy and F. Al-Kharafi, The inhibition of the corrosion of Al, Al-6061 and Al-Cu in chloride free aqueous media: I. Passivation in acid solutions. *Corrosion Science*, 39(4), pp. 681–700, 1997. DOI: 10.1016/s0010-938x(97)89336-5. 1

[38] H. Wang and R. Akid, Encapsulated cerium nitrate inhibitors to provide high-performance anti-corrosion sol—gel coatings on mild steel. *Corrosion Science*, 50(4), pp. 1142–1148, 2008. DOI: 10.1016/j.corsci.2007.11.019. 1

[39] R. Saleh, A. Ismail, and A. El Hosary, Corrosion inhibition by naturally occurring substances-IX. The effect of the aqueous extracts of some seeds, leaves, fruits and fruit-peels on the corrosion of Al in NaOH. *Corrosion Science*, 23(11), pp. 1239–1241, 1983. DOI: 10.1016/0010-938x(83)90051-3. 2

[40] F. Zucchi and I. H. Omar, Plant extracts as corrosion inhibitors of mild steel in HCl solutions. *Surface Technology*, 24(4), pp. 391–399, 1985. DOI: 10.1016/0376-4583(85)90057-3. 2

[41] S. Deng and X. Li, Inhibition by Ginkgo leaves extract of the corrosion of steel in HCl and H2SO4 solutions. *Corrosion Science*, 55, pp. 407–415, 2012. DOI: 10.1016/j.corsci.2011.11.005. 2, 5, 28, 30

[42] O. K. Abiola and A. James, The effects of Aloe vera extract on corrosion and kinetics of corrosion process of zinc in HCl solution. *Corrosion Science*, 52(2), pp. 661–664, 2010. DOI: 10.1016/j.corsci.2009.10.026. 2

[43] I. Farooqi, M. Quraishi, and P. Saini, Corrosion prevention of mild steel in 3% NaCl water by some naturally-occurring substances. *Corrosion Prevention and Control*, 46(4), pp. 93–96, 1999. 2

[44] O. K. Abiola, J. Otaigbe, and O. Kio, Gossipium hirsutum L. extracts as green corrosion inhibitor for aluminum in NaOH solution. *Corrosion Science*, 51(8), pp. 1879–1881, 2009. DOI: 10.1016/j.corsci.2009.04.016. 4

[45] A. Abdel-Gaber, et al., Inhibitive action of some plant extracts on the corrosion of steel in acidic media. *Corrosion Science*, 48(9), pp. 2765–2779, 2006. DOI: 10.1016/j.corsci.2005.09.017.

[46] L. Valek and S. Martinez, Copper corrosion inhibition by Azadirachta indica leaves extract in 0.5 M sulphuric acid. *Materials Letters*, 61(1), pp. 148–151, 2007. DOI: 10.1016/j.matlet.2006.04.024.

[47] X. Li, S. Deng, and H. Fu, Inhibition of the corrosion of steel in HCl, H2SO4 solutions by bamboo leaf extract. *Corrosion Science*, 62, pp. 163–175, 2012. DOI: 10.1016/j.corsci.2012.05.008. 2

[48] M. M. Fares, A. Maayta, and M. M. Al-Qudah, Pectin as promising green corrosion inhibitor of aluminum in hydrochloric acid solution. *Corrosion Science*, 60, pp. 112–117, 2012. DOI: 10.1016/j.corsci.2012.04.002. 2

[49] Y. Ren, et al., Lignin terpolymer for corrosion inhibition of mild steel in 10% hydrochloric acid medium. *Corrosion Science*, 50(11), pp. 3147–3153, 2008. DOI: 10.1016/j.corsci.2008.08.019. 2

[50] L. Chauhan and G. Gunasekaran, Corrosion inhibition of mild steel by plant extract in dilute HCl medium. *Corrosion Science*, 49(3), pp. 1143–1161, 2007. DOI: 10.1016/j.corsci.2006.08.012. 2

[51] A. El-Etre, M. Abdallah, and Z. El-Tantawy, Corrosion inhibition of some metals using lawsonia extract. *Corrosion Science*, 47(2), pp. 385–395, 2005. DOI: 10.1016/j.corsci.2004.06.006. 2

[52] R. Saleh and A. Ismail, Corrosion inhibition by naturally-occurring substances: The effect of fenugreek, lupine, doum, beet and solanum melongena extracts on the corrosion of steel, Al, Zn and Cu in acids. *Corrosion Prevention and Control*, 31, p. 21, 1984. 2

[53] A. El Hosary, R. Saleh, and A. S. El Din, Corrosion inhibition by naturally occurringsubstances—I. The effect of Hibiscus subdariffa (karkade) extract on the dissolution of Al and Zn. *Corrosion Science*, 12(12), pp. 897–904, 1972. DOI: 10.1016/s0010-938x(72)80098-2. 2

[54] M. M. Fares, A. Maayta, and J. A. Al-Mustafa, Corrosion inhibition of iota-carrageenan natural polymer on aluminum in presence of zwitterion mediator in HCl media. *Corrosion Science*, 65, pp. 223–230, 2012. DOI: 10.1016/j.corsci.2012.08.018. 2, 57, 60

[55] E. Oguzie, et al., Adsorption and corrosion-inhibiting effect of Dacryodis edulis extract on low-carbon-steel corrosion in acidic media. *Journal of Colloid and Interface Science*, 349(1), pp. 283–292, 2010. DOI: 10.1016/j.jcis.2010.05.027. 2, 4, 5, 6, 27, 30

[56] P. B. Raja and M. G. Sethuraman, Natural products as corrosion inhibitor for metals in corrosive media—a review. *Materials Letters*, 62(1), pp. 113–116, 2008. DOI: 10.1016/j.matlet.2007.04.079. 2, 3

[57] M. Finšgar and J. Jackson, Application of corrosion inhibitors for steels in acidic media for the oil and gas industry: A review. *Corrosion Science*, 86, pp. 17–41, 2014. DOI: 10.1016/j.corsci.2014.04.044. 3, 6, 9

[58] C. G. Dariva and A. F. Galio, Corrosion inhibitors—principles, mechanisms and applications. *Developments in Corrosion Protection*, InTech, 2014. DOI: 10.5772/57255. 3

[59] G. Ji, et al., Inhibitive effect of chlorophytum borivilianum root extract on mild steel corrosion in HCl and H2SO4 solutions. *Industrial and Engineering Chemistry Research*, 52(31), pp. 10673–10681, 2013. DOI: 10.1021/ie4008387. 3, 6, 27, 30

[60] A. M. Abdel-Gaber, B. A. Abd-El-Nabey, and M. Saadawy, The role of acid anion on the inhibition of the acidic corrosion of steel by lupine extract. *Corrosion Science*, 51(5), pp. 1038–1042, 2009. DOI: 10.1016/j.corsci.2009.03.003. 4

[61] M. Quraishi, et al., Green approach to corrosion inhibition of mild steel in hydrochloric acid and sulphuric acid solutions by the extract of Murraya koenigii leaves. *Materials Chemistry and Physics*, 122(1), pp. 114–122, 2010. DOI: 10.1016/j.matchemphys.2010.02.066. 4

[62] M. Behpour, et al., Green approach to corrosion inhibition of mild steel in two acidic solutions by the extract of Punica granatum peel and main constituents. *Materials Chemistry and Physics*, 131(3), pp. 621–633, 2012. DOI: 10.1016/j.matchemphys.2011.10.027. 4, 6, 28, 31

[63] A. Abdel-Gaber, et al., Inhibition of aluminium corrosion in alkaline solutions using natural compound. *Materials Chemistry and Physics*, 109(2–3), pp. 297-305, 2008. DOI: 10.1016/j.matchemphys.2007.11.038. 4, 36, 38

[64] M. Lebrini, et al., Corrosion inhibition of C38 steel in 1 M hydrochloric acid medium by alkaloids extract from Oxandra asbeckii plant. *Corrosion Science*, 53(2), pp. 687–695, 2011. DOI: 10.1016/j.corsci.2010.10.006. 4

[65] P. Okafor, et al., Inhibitory action of Phyllanthus amarus extracts on the corrosion of mild steel in acidic media. *Corrosion Science*, 50(8), pp. 2310–2317, 2008. DOI: 10.1016/j.corsci.2008.05.009. 4

[66] K. V. Kumar, M. S. N. Pillai, and G. R. Thusnavis, Seed extract of Psidium guajava as eco-friendly corrosion inhibitor for carbon steel in hydrochloric acid medium. *Journal of Materials Science and Technology*, 27(12), pp. 1143–1149, 2011. DOI: 10.1016/s1005-0302(12)60010-3. 4

[67] H. Z. Alkhathlan, et al., Launaea nudicaulis as a source of new and efficient green corrosion inhibitor for mild steel in acidic medium: a comparative study of two solvent extracts. *International Journal of Electrochemical Science*, 9, pp. 870–89, 2014. 4, 12, 20

[68] S. Garai, et al., A comprehensive study on crude methanolic extract of Artemisia pallens (Asteraceae) and its active component as effective corrosion inhibitors of mild steel in acid solution. *Corrosion Science*, 60, pp. 193–204, 2012. DOI: 10.1016/j.corsci.2012.03.036. 4

[69] H. Gerengi, H. Goksu, and P. Slepski, The inhibition effect of mad honey on corrosion of 2007-type aluminium alloy in 3.5% NaCl solution. *Materials Research*, 17(1), pp. 255–264, 2014. DOI: 10.1590/s1516-14392013005000174. 4, 29, 34

[70] E. E. Oguzie, Evaluation of the inhibitive effect of some plant extracts on the acid corrosion of mild steel. *Corrosion Science*, 50(11), pp. 2993–2998, 2008. DOI: 10.1016/j.corsci.2008.08.004. 4, 27, 30

[71] M. A. Chidiebere, et al., Corrosion inhibition and adsorption behavior of Punica granatum extract on mild steel in acidic environments: Experimental and theoretical studies. *Industrial and Engineering Chemistry Research*, 51(2), pp. 668–677, 2012. DOI: 10.1021/ie201941f. 5, 6, 28, 30

[72] A. S. Yaro, A. A. Khadom, and R. K. Wael, Apricot juice as green corrosion inhibitor of mild steel in phosphoric acid. *Alexandria Engineering Journal*, 52(1), pp. 129–135, 2013. DOI: 10.1016/j.aej.2012.11.001. 5, 44, 46

[73] M. Abdallah, et al., Natural oils as corrosion inhibitors for stainless steel in sodium hydroxide solutions. *Chemistry and Technology of Fuels and Oils*, 48(3), pp. 234–245, 2012. DOI: 10.1007/s10553-012-0364-x. 6, 52, 55

[74] M. Behpour, et al., Inhibition of 304 stainless steel corrosion in acidic solution by Ferula gumosa (galbanum) extract. *Materials and Corrosion*, 60(11), pp. 895–898, 2009. DOI: 10.1002/maco.200905182. 10, 18

[75] N. Soltani, et al., Green approach to corrosion inhibition of 304 stainless steel in hydrochloric acid solution by the extract of Salvia officinalis leaves. *Corrosion Science*, 62, pp. 122–135, 2012. DOI: 10.1016/j.corsci.2012.05.003. 10, 19

[76] R. Prabhu, T. Venkatesha, and A. Shanbhag, Carmine and fast green as corrosion in-hibitors for mild steel in hydrochloric acid solution. *Journal of the Iranian Chemical Society*, 6(2), pp. 353–363, 2009. DOI: 10.1007/bf03245845. 10, 19

[77] P. Nagarajan, et al., Natural product extract as eco-friendly corrosion inhibitor for commercial mild steel in 1 M HCI-Part II. *Journal of Indian Council of Chemists*, 26(2), pp. 53–157, 2009. 10, 19

[78] A. Sultan, et al., Study of some natural products as eco-friendly corrosion inhibitor for mild steel in 1.0 M HCl solution. 2013. 11, 19

[79] M. H. Hussin and M. J. Kassim, Electrochemical, thermodynamic and adsorption studies of (+)-catechin hydrate as natural mild steel corrosion inhibitor in 1 M HCl. *International Journal of Electrochemical Science*, 6(5), pp. 1396–1414, 2011. 11, 19

[80] M. Mobin and M. Rizvi, Inhibitory effect of xanthan gum and synergistic surfactant additives for mild steel corrosion in 1 M HCl. *Carbohydrate Polymers*, 136, pp. 384–393, 2016. DOI: 10.1016/j.carbpol.2015.09.027. 11, 19

[81] A. Alobaidy, et al., Eco-friendly corrosion inhibitor: Experimental studies on the corrosion inhibition performance of creatinine for mild steel in HCl complemented with quantum chemical calculations. *International Journal of Electrochemical Science*, 10, pp. 3961–3972, 2015. 11, 20

[82] T. F. Souza, et al., Inhibitory action of ilex paraguariensis extracts on the corrosion of carbon steel in HCl solution. *International Journal of Electrochemical Science*, 10(1), pp. 22–33, 2015. 11, 20

[83] M. A. Chidiebere, et al., Inhibitory action of funtumia elastica extracts on the corrosion of Q235 mild steel in hydrochloric acid medium: Experimental and theoretical studies. *Journal of Dispersion Science and Technology*, 36(8), pp. 1115–1125, 2015. DOI: 10.1080/01932691.2014.956114. 12, 20

[84] H. Alkhathlan, et al., Anticorrosive assay-guided isolation of active phytoconstituents from Anthemis pseudocotula extracts and a detailed study of their effects on the corrosion of mild steel in acidic media. *RSC Advances*, 5(67), pp. 54283–54292, 2015. DOI: 10.1039/c5ra09154c. 12, 20

[85] R. Prabhu, et al., Inhibition effect of Azadirachta indica, a natural product, on the corrosion of zinc in hydrochloric acid solution. *Transactions of the Indian Institute of Metals*, 67(5), pp. 675–679, 2014. DOI: 10.1007/s12666-014-0390-y. 12, 20

[86] A. M. Atta, et al., Corrosion inhibition of mild steel in acidic medium by magnetite myrrh nanocomposite. *International Journal of Electrochemical Science*, 9, pp. 8446–8457, 2014. 13, 21

[87] L. Li, et al., Experimental and theoretical investigations of Michelia alba leaves extract as a green highly-effective corrosion inhibitor for different steel materials in acidic solution. *RSC Advances*, 5(114), pp. 93724–93732, 2015. DOI: 10.1039/c5ra19088f. 13, 21

[88] M. Belkhaouda, et al., Inhibition of C-steel corrosion in hydrochloric solution with chenopodium ambrorsioides extract. *International Journal of Electrochemical Science*, 8, pp. 7425–7436, 2013. 13, 21

[89] M. Nasibi, et al., Chamomile (Matricaria recutita) extract as a corrosion inhibitor for mild steel in hydrochloric acid solution. *Chemical Engineering Communications*, 200(3), pp. 367–378, 2013. DOI: 10.1080/00986445.2012.709475. 13, 21

[90] L. Bammou, et al., Inhibition effect of natural junipers extract towards steel corrosion in HCl solution. *International Journal of Electrochemical Science*, 7, pp. 8974–8987, 2012. 13, 21

[91] A. Shah, et al., Green inhibitors for copper corrosion by mangrove tannin. *International Journal of Electrochemical Science*, 8(2), pp. 2140–2153, 2013. 13, 21

[92] L. Afia, et al., Comparative study of corrosion inhibition on mild steel in HCl medium by three green compounds: Argania spinosa press cake, kernels and hulls extracts. *Transactions of the Indian Institute of Metals*, 66(1), pp. 43–49, 2013. DOI: 10.1007/s12666-012-0168-z. 14, 22

[93] T. Ibrahim, H. Alayan, and Y. Al Mowaqet, The effect of thyme leaves extract on corrosion of mild steel in HCl. *Progress in Organic Coatings*, 75(4), pp. 456–462, 2012. DOI: 10.1016/j.porgcoat.2012.06.009. 14, 22

[94] L. Afia, et al., Testing natural compounds: Argania spinosa kernels extract and cosmetic oil as eco-friendly inhibitors for steel corrosion in 1 M HCl. *International Journal of Electrochemical Science*, 6, pp. 5918–5939, 2011. 14, 22

[95] M. Mahat, et al., Azadirachta excelsa as green corrosion inhibitor for mild steel in acidic medium. *Business, Engineering and Industrial Applications (ISBEIA)*, Symposium on IEEE, 2012. DOI: 10.1109/isbeia.2012.6422944. 14, 22

[96] M. Dahmani, et al., Corrosion inhibition of C38 steel in 1 M HCl: A comparative study of black pepper extract and its isolated pipeline. *International Journal of Electrochemical Science*, 5(8), pp. 1060–1069, 2010. 14, 22

[97] V. V. Torres, et al., Inhibitory action of aqueous coffee ground extracts on the corrosion of carbon steel in HCl solution. *Corrosion Science*, 53(7), pp. 2385–2392, 2011. DOI: 10.1016/j.corsci.2011.03.021. 14, 22

[98] S. Subhashini, et al., Corrosion mitigating effect of Cyamopsis Tetragonaloba seed extract on mild steel in acid medium. *Journal of Chemistry*, 7(4), pp. 1133–1137, 2010. DOI: 10.1155/2010/457825. 15, 22

[99] T. Ibrahim and M. Habbab, Corrosion inhibition of mild steel in 2 M HCl using aqueous extract of eggplant peel. *International Journal of Electrochemical Science*, 6, pp. 5357–5371, 2011. 15, 22

[100] T. H. Ibrahim and M. A. Zour, Corrosion inhibition of mild steel using fig leaves extract in hydrochloric acid solution. *International Journal of Electrochemical Science*, 6(12), pp. 6442–6455, 2011. 15, 23

[101] R. Goel, et al., Corrosion inhibition of mild steel in HCl by isolated compounds of Riccinus communis (L.). *Journal of Chemistry*, 7(S1), pp. S319-S329, 2010. DOI: 10.1155/2010/308057. 15, 23

[102] M. Dahmani, et al., Investigation of piperanine as HCl eco-friendly corrosion inhibitors for C38 steel. *International Journal of Electrochemical Science*, 7, pp. 2513–2522, 2012. 15, 23

[103] T. H. Ibrahim, Y. Chehade, and M. A. Zour, Corrosion inhibition of mild steel using potato peel extract in 2 M HCL solution. *International Journal of Electrochemical Science*, 6(12), pp. 6542–6555, 2011. 15, 23

[104] Y. Abboud, et al., Corrosion inhibition of carbon steel in acidic media by Bifurcaria bifurcata extract. *Chemical Engineering Communications*, 196(7), pp. 788–800, 2009. DOI: 10.1080/00986440802589875. 15, 23

[105] A. Bouyanzer, et al., Testing natural fenugreek as an eco-friendly inhibitor for steel corrosion in 1 M HCl. *Portugaliae Electrochimica Acta*, 28(3), pp. 165–172, 2010. DOI: 10.4152/pea.201003165. 16, 23

[106] I. Obot and N. Obi-Egbedi, Ginseng root: A new efficient and effective eco-friendly corrosion inhibitor for aluminium alloy of type AA 1060 in hydrochloric acid solution. *International Journal of Electrochemical Science*, 4(9), pp. 1277–1288, 2009. 16, 24

[107] A. Satapathy, et al., Corrosion inhibition by Justicia gendarussa plant extract in hydrochloric acid solution. *Corrosion Science*, 51(12), pp. 2848–2856, 2009. DOI: 10.1016/j.corsci.2009.08.016. 16, 24

[108] S. Umoren and E. Ebenso, Studies of the anti-corrosive effect of Raphia hookeri exudate gum-halide mixtures for aluminium corrosion in acidic medium. *Pigment and Resin Technology*, 37(3), pp. 173–182, 2008. DOI: 10.1108/03699420810871020. 16, 24

[109] A. El-Etre, Khillah extract as inhibitor for acid corrosion of SX 316 steel. *Applied Surface Science*, 252(24), pp. 8521–8525, 2006. DOI: 10.1016/j.apsusc.2005.11.066. 16, 24

[110] S. Mahmoud, Corrosion inhibition of muntz (63% Cu, » 37% Zn) alloy in HCl solution by some naturally occurring extracts. *Portugaliae Electrochimica Acta*, 24(4), pp. 441–455, 2006. DOI: 10.4152/pea.200604441. 17, 25

[111] R. S. Mayanglambam, V. Sharma, and G. Singh, Musa paradisiaca extract as a green inhibitor for corrosion of mild steel in 0.5 M sulphuric acid solution. *Portugaliae Electrochimica Acta*, 29(6), pp. 405–417, 2011. DOI: 10.4152/pea.201106405. 17, 26

[112] P. B. Raja and M. Sethuraman, Solanum nigrum as natural source of corrosion inhibitor for mild steel in sulphuric acid medium. *Corrosion Engineering, Science and Technology*, 45(6), pp. 455–460, 2010. DOI: 10.1179/147842208x388762. 17, 26

[113] N. Eddy and E. Ebenso, Corrosion inhibition and adsorption properties of ethanol extract of Gongronema latifolium on mild steel in H2SO4. *Pigment and Resin Technology*, 39(2), pp. 77–83, 2010. DOI: 10.1108/03699421011028653. 17, 26

[114] M. Bouklah and B. Hammouti, Thermodynamic characterisation of steel corrosion for the corrosion inhibition of steel in sulphuric acid solutions by Artemisia. *Portugaliae Electrochimica Acta*, 24(4), pp. 457–468, 2006. DOI: 10.4152/pea.200604457. 25, 26

[115] L. Bammou, et al., Corrosion inhibition of steel in sulfuric acidic solution by the Chenopodium Ambrosioides Extracts. *Journal of the Association of Arab Universities for Basic and Applied Sciences*, 16, pp. 83–90, 2014. DOI: 10.1016/j.jaubas.2013.11.001. 25, 26

[116] B. Anand and V. Balasubramanian, Study on corrosion inhibition of mild steel using natural product as corrosion inhibitor in acidic medium. 2011. 28, 31

[117] A. Alsabagh, et al., Utilization of green tea as environmentally friendly corrosion inhibitor for carbon steel in acidic media. *International Journal of Electrochemical Science*, 10, pp. 1855–1872, 2015. 28, 31

[118] E. E. Oguzie, et al., Natural products for materials protection: Mechanism of corrosion inhibition of mild steel by acid extracts of Piper guineense. *The Journal of Physical Chemistry C*, 116(25), pp. 13603–13615, 2012. DOI: 10.1021/jp300791s. 28, 31

[119] A. A. A. Hadi, Use Reed leaves as a natural inhibitor to reduce the corrosion of low carbon steel. 29, 34

[120] A. El-Etre, Natural honey as corrosion inhibitor for metals and alloys. I. Copper in neutral aqueous solution. *Corrosion Science*, 40(11), pp. 1845–1850, 1998. DOI: 10.1016/s0010-938x(98)00082-1. 29, 34

[121] A. Singh, et al., Application of a natural inhibitor for corrosion inhibition of J55 steel in CO2 saturated 3.5% NaCl solution. *International Journal of Electrochemical Science*, 8, pp. 12851–12859, 2013. 29, 34

[122] W. Liu, et al., Corrosion inhibition of Al-alloy in 3.5% NaCl solution by a natural inhibitor: An electrochemical and surface study. *International Journal of Electrochemical Science*, 9, pp. 5560–5573, 2014. 29, 34

[123] A. A. Al-Asadi, et al., Effect of an Aloe Vera as a natural inhibitor on the corrosion of mild steel in 1 wt. % NaCl. *International Research Journal of Engineering and Technology (IRJET)*, 2(6), 2015. 29, 34

[124] M. Shabani-Nooshabadi and M. Ghandchi, Santolina chamaecyparissus extract as a natural source inhibitor for 304 stainless steel corrosion in 3.5% NaCl. *Journal of Industrial and Engineering Chemistry*, 31, pp. 231–237, 2015. DOI: 10.1016/j.jiec.2015.06.028. 32, 34

[125] J. O. Okeniyi, C. A. Loto, and A. P. I. Popoola, Investigating the corrosion mechanism of Morinda lucida leaf extract admixtures on concrete steel rebar in saline/marine simulating environment. *International Journal of Electrochemical Science*, 10(12), pp. 9893–9906, 2015. 32, 34

[126] B. Abd-El-Naby, et al., Effect of some natural extracts on the corrosion of zinc in 0.5 M NaCl. *International Journal of Electrochemical Science*, 7, pp. 5864–5879, 2012. 32, 34

[127] I. Radojčić, et al., Natural honey and black radish juice as tin corrosion inhibitors. *Corrosion Science*, 50(5), pp. 1498–1504, 2008. DOI: 10.1016/j.corsci.2008.01.013. 32, 35

[128] R. B. Channouf, N. Souissi, and N. Bellakhal, Juniperus communis extract effect on bronze corrosion in natural 0.5 M chloride medium. *Journal of Materials Science and Chemical Engineering*, 3(11), p. 21, 2015. DOI: 10.4236/msce.2015.311004. 32, 35

[129] A. El-Etre and M. Abdallah, Natural honey as corrosion inhibitor for metals and alloys. II. C-steel in high saline water. *Corrosion Science*, 42(4), pp. 731–738, 2000. DOI: 10.1016/s0010-938x(99)00106-7. 33, 37

[130] W. W. Nik, et al., Study of henna (Lawsonia inermis) as natural corrosion inhibitor for aluminum alloy in seawater. *IOP Conference Series: Materials Science and Engineering*, IOP Publishing, 2012. DOI: 10.1088/1757-899x/36/1/012043. 33, 37

[131] M. Felipe, et al., Effectiveness of Croton cajucara benth on corrosion inhibition of carbon steel in saline medium. *Materials and Corrosion*, 64(6), pp. 530–534, 2013. DOI: 10.1002/maco.201206532. 33, 37

[132] F. Rachmanda, et al., Corrosion behavior of API-5L in various green inhibitors. *Advanced Materials Research*, Transactions on Tech Publ., 2013. DOI: 10.4028/www.scientific.net/AMR.634-638.689. 33, 37

[133] J. Buchweishaija and G. Mhinzi, Natural products as a source of environmentally friendly corrosion inhibitors: The case of gum exudate from Acacia seyal var. seyal. *Portugaliae Electrochimica Acta*, 26(3), pp. 257–265, 2008. DOI: 10.4152/pea.2008032257. 33, 37

[134] A. Buyuksagis, M. Dilek, and M. Kargioglu, Corrosion inhibition of st37 steel in geothermal fluid by Quercus robur and pomegranate peels extracts. *Protection of Metals and Physical Chemistry of Surfaces*, 51(5), pp. 861–872, 2015. DOI: 10.1134/s2070205115050056. 33, 38

[135] T. Ramde, S. Rossi, and C. Zanella, Inhibition of the Cu65/Zn35 brass corrosion by natural extract of Camellia sinensis. *Applied Surface Science*, 307, pp. 209–216, 2014. DOI: 10.1016/j.apsusc.2014.04.016. 35, 38

[136] A. Motalebi, et al., Improvement of corrosion performance of 316L stainless steel via PVTMS/henna thin film. *Progress in Natural Science: Materials International*, 22(5), pp. 392–400, 2012. DOI: 10.1016/j.pnsc.2012.10.006. 35, 38

[137] A. Abdel-Gaber, B. Abdel-Nabey, and M. Saadawy, The co-operative effect of chloride ions and some natural extracts in retarding corrosion of steel in neutral media. *Materials and Corrosion*, 63(2), pp. 161–167, 2012. DOI: 10.1002/maco.201005678. 36, 38

[138] M. Abdallah, et al., Natural occurring substances as corrosion inhibitors for tin in sodium bicarbonate solutions. *Journal of the Korean Chemical Society*, 53(4), pp. 485–490, 2009. DOI: 10.5012/jkcs.2009.53.5.485. 36, 38

[139] A. A. Torres-Acosta, Opuntia-Ficus-Indica (Nopal) mucilage as a steel corrosion inhibitor in alkaline media. *Journal of Applied Electrochemistry*, 37(7), pp. 835–841, 2007. DOI: 10.1007/s10800-007-9319-z. 36, 39

[140] J. Flores-De los Rios, et al., Opuntia ficus-indica extract as green corrosion inhibitor for carbon steel in 1 M HCl solution. *Journal of Spectroscopy*, 2015. DOI: 10.1155/2015/714692. 41, 42

[141] J. Zhang, et al., Investigation of Diospyros Kaki Lf husk extracts as corrosion inhibitors and bactericide in oil field. *Chemistry Central Journal*, 7(1), pp. 109, 2013. DOI: 10.1186/1752-153x-7-109. 41, 42

[142] J. A. D. C. P. Gomes, J. C. Rocha, and E. D'elia, Use of fruit skin extracts as corrosion inhibitors and process for producing same. Google Patents, 2015. 41, 42

[143] P. Matheswaran, et al., Corrosion inhibition on mild steel in sulphuric acid medium using natural product as inhibitor. DOI: 10.7598/cst2015.961. 43, 45

[144] S. M. Mahdi, Study the pomegranate's peel powder as a natural inhibitor for mild steel corrosion. 43, 45

[145] N. Odewunmi, S. Umoren, and Z. Gasem, Utilization of watermelon rind extract as a green corrosion inhibitor for mild steel in acidic media. *Journal of Industrial and Engineering Chemistry*, 21, pp. 239–247, 2015. DOI: 10.1016/j.jiec.2014.02.030. 43, 45

[146] S. A. Umoren, Z. M. Gasem, and I. B. Obot, Natural products for material protection: Inhibition of mild steel corrosion by date palm seed extracts in acidic media. *Industrial and Engineering Chemistry Research*, 52(42), pp. 14855–14865, 2013. DOI: 10.1021/ie401737u. 43, 45

[147] C. Loganayagi, C. Kamal, and M. Sethuraman, Opuntiol: An active principle of opuntia elatior as an eco-friendly inhibitor of corrosion of mild steel in acid medium. *ACS Sustainable Chemistry and Engineering*, 2(4), pp. 606–613, 2014. DOI: 10.1021/sc4003642. 43, 45

[148] M. Abdulwahab, et al., Effect of Avogadro natural oil on the corrosion inhibition of mild steel in hydrochloric acid solution. *Research on Chemical Intermediates*, 40(3), pp. 1115–1123, 2014. DOI: 10.1007/s11164-013-1025-3. 47, 49

[149] L. Bammou, et al., Inhibition effect of natural Artemisia oils towards tinplate corrosion in HCL solution: Chemical characterization and electrochemical study. *International Journal of Electrochemical Science*, 6, pp. 1454–1467, 2011. 47, 49

[150] N. Lahhit, et al., Fennel (Foeniculum vulgare) essential oil as green corrosion inhibitor of carbon steel in hydrochloric acid solution. *Portugaliae Electrochimica Acta*, 29(2), pp. 127–138, 2011. DOI: 10.4152/pea.201102127. 47, 49

[151] M. Znini, et al., Chemical composition and inhibitory effect of Mentha spicata essential oil on the corrosion of steel in molar hydrochloric acid. *International Journal of Electrochemical Science*, 6(3), pp. 691–704, 2011. 48, 49

[152] M. Abdallah, S. Al Karanee, and A. Abdel Fatah, Inhibition of acidic and pitting corrosion of nickel using natural black cumin oil. *Chemical Engineering Communications*, 197(12), pp. 1446–1454, 2010. DOI: 10.1080/00986445.2010.484982. 48, 49

[153] L. Bammou, et al., Thermodynamic properties of Thymus satureioides essential oils as corrosion inhibitor of tinplate in 0.5 M HCl: Chemical characterization and electrochemical study. *Green Chemistry Letters and Reviews*, 3(3), pp. 173–178, 2010. DOI: 10.1080/17518251003660121. 48, 49

[154] A. Bouyanzer, B. Hammouti, and L. Majidi, Pennyroyal oil from Mentha pulegium as corrosion inhibitor for steel in 1 M HCl. *Materials Letters*, 60(23), pp. 2840–2843, 2006. DOI: 10.1016/j.matlet.2006.01.103. 48, 50

[155] M. Znini, et al., Essential oil of Salvia aucheri mesatlantica as a green inhibitor for the corrosion of steel in 0.5 MH 2 SO 4. *Arabian Journal of Chemistry*, 5(4), pp. 467–474, 2012. DOI: 10.1016/j.arabjc.2010.09.017. 48, 51

[156] S. Rekkab, et al., Green corrosion inhibitor from essential oil of Eucalyptus globulus (Myrtaceae) for C38 steel in sulfuric acid solution. *Journal of Materials and Environmental Science*, 3(4), pp. 613–627, 2012. 48, 51

[157] O. Ouachikh, et al., Application of essential oil of Artemisia herba alba as green corrosion inhibitor for steel in 0.5 MH 2 SO 4. *Surface Review and Letters*, 16(01), pp. 49–54, 2009. DOI: 10.1142/s0218625x09012287. 50, 51

[158] O. Fayomi and A. Popoola, The inhibitory effect and adsorption mechanism of roasted Elaeis guineensis as green inhibitor on the corrosion processor extruded AA6063 Al-Mg-Si alloy in simulated solution. *Silicon*, 6(2), pp. 137–143, 2014. DOI: 10.1007/s12633-014-9177-3. 50, 53

[159] M. Abdulwahab, et al., Effect of Avogadro oil as corrosion inhibitor of thermally pre-aged Al-Si-Mg alloy in sodium chloride solution. *Silicon*, 5(3), pp. 225–228, 2013. DOI: 10.1007/s12633-013-9156-0. 50, 53

[160] A. Abdel Nazeer, et al., Effect of glycine on the electrochemical and stress corrosion cracking behavior of Cu10Ni alloy in sulfide polluted salt water. *Industrial and Engineering Chemistry Research*, 50(14), pp. 8796–8802, 2011. DOI: 10.1021/ie200763b. 50, 53

[161] J. Halambek, K. Berković, and J. Vorkapić-Furač, The influence of Lavandula angustifolia L. oil on corrosion of Al-3Mg alloy. *Corrosion Science*, 52(12), pp. 3978–3983, 2010. DOI: 10.1016/j.corsci.2010.08.012. 52, 53

[162] J. Porcayo-Calderon, et al., Imidazoline derivatives based on coffee oil as CO2 corrosion inhibitor. *International Journal of Electrochemical Science*, 10, pp. 3160–3176, 2015. 52, 53

[163] M. Abdallah, et al., Corrosion behavior of nickel electrode in NaOH solution and its inhibition by some natural oils. *International Journal of Electrochemical Science*, 9, pp. 1071–1086, 2014. 52, 55

[164] M. Abdallah, et al., Use of some natural oils as crude pipeline corrosion inhibitors in sodium hydroxide solutions. *Chemistry and Technology of Fuels and Oils*, 46(5), pp. 354–362, 2010. DOI: 10.1007/s10553-010-0234-3. 54, 55

[165] S. Houbairi, M. Essahli, and A. Lamiri, A natural extract as corrosion inhibitor for copper surface in acid solution. *International Journal of Engineering Research and Technology*, ESRSA Publications, 2014. 54, 55

[166] S. Houbairi, M. Essahli, and A. Lamiri, Inhibition of copper corrosion in 2 M HNO3 by the essential oil of thyme morocco. *Portugaliae Electrochimica Acta*, 31(4), pp. 221–233, 2013. DOI: 10.4152/pea.201304221. 54, 55

[167] M. Benabdellah, et al., Inhibition of steel corrosion in 2 M H3PO4 by artemisia oil. *Applied Surface Science*, 252(18), pp. 6212–6217, 2006. DOI: 10.1016/j.apsusc.2005.08.030. 54, 56

[168] M. Bendahou, M. Benabdellah, and B. Hammouti, A study of rosemary oil as a green corrosion inhibitor for steel in 2 M H3PO4. *Pigment and Resin Technology*, 35(2), pp. 95–100, 2006. DOI: 10.1108/03699420610652386. 54, 56

[169] G. Shahi, et al., Thermodynamic and electrochemical investigation of (9-[(R) 2 [[bis [[(isopropoxycarbonyl) oxy] methoxy] phosphinyl] methoxy] propyl] adenine fumarate) as green corrosion inhibitor for mild steel in 1 M HCl. *International Journal of Electrochemical Science*, 10, pp. 1102–1116, 2015. 57, 60

[170] S. A. Umoren, et al., Inhibition of mild steel corrosion in HCl solution using chitosan. *Cellulose*, 20(5), pp. 2529–2545, 2013. DOI: 10.1007/s10570-013-0021-5. 57, 60

[171] C. Verma, M. Quraishi, and E. Ebenso, Thermodynamics and electrochemical investigation of (4-(2-amino-6-(cyclopropylamino)-9H-purin-9-yl) cyclopent-2-enyl) methanol sulphate as green and effective corrosion inhibitor for mild steel in 1 M hydrochloric acid. *International Journal of Electrochemical Science*, 8, pp. 12238–12251, 2013. 57, 60

[172] A. Fouda, A. El-desoky, and H. M. Hassan, Quinazoline derivatives as green corrosion inhibitors for carbon steel in hydrochloric acid solutions. *International Journal of Electrochemical Science*, 8, pp. 5866–5885, 2013. 58, 60

[173] J. Da Rocha, et al., Grape Pomace extracts as green corrosion inhibitors for carbon steel in hydrochloric acid solutions. *International Journal of Electrochemical Science*, 7(12), pp. 11941–11956, 2012. 58, 61

[174] P. Premkumar, K. Kannan, and M. Natesan, Effect of menthol coated craft paper on corrosion of copper in HCl environment. *Bulletin of Materials Science*, 33(3), pp. 307–311, 2010. DOI: 10.1007/s12034-010-0047-3. 58, 61

[175] Z. Faska, et al., Effect of pulegone and pulegone oxide on the corrosion of steel in 1 M HCl. *Monatshefte Für Chemie/Chemical Monthly*, 139(12), pp. 1417–1422, 2008. DOI: 10.1007/s00706-008-0959-4. 58, 61

[176] E. Chaieb, et al., Inhibition of the corrosion of steel in 1 M HCl by eugenol derivatives. *Applied Surface Science*, 246(1), pp. 199–206, 2005. DOI: 10.1016/j.apsusc.2004.11.011. 58, 61

[177] M. I. Awad, Eco-friendly corrosion inhibitors: Inhibitive action of quinine for corrosion of low carbon steel in 1 M HCl. *Journal of Applied Electrochemistry*, 36(10), pp. 1163–1168, 2006. DOI: 10.1007/s10800-006-9204-1. 59, 61

[178] A. A. Nazeer, H. El-Abbasy, and A. Fouda, Antibacterial drugs as environmentally-friendly corrosion inhibitors for carbon steel in acid medium. *Research on Chemical Intermediates*, 39(3), pp. 921–939, 2013. DOI: 10.1007/s11164-012-0605-y. 59, 62

[179] A. Popoola, M. Abdulwahab, and O. Fayomi, Corrosion inhibition of mild steel in Sesamum indicum-2M HCl/H2SO4 interface. *International Journal of Electrochemical Science*, 7(7), pp. 5805–5816, 2012. 59, 62

[180] A. Fouda, A. El-Defrawy, and M. El-Sherbeni, Lornoxicam and Tenoxicam drugs as green corrosion inhibitors for carbon steel in 1 MH 2 SO 4 solution. *Journal of Electrochemical Science and Technology*, 4(2), pp. 47–56, 2013. DOI: 10.5229/jecst.2013.4.2.47. 59, 62

[181] E. Akbarzadeh, M. M. Ibrahim, and A. A. Rahim, Corrosion inhibition of mild steel in near neutral solution by kraft and soda lignins extracted from oil palm empty fruit bunch. *International Journal of Electrochemical Science*, 6(11), pp. 5396–5416, 2011. 59, 64

[182] R. Solmaz, et al., Citric acid as natural corrosion inhibitor for aluminium protection. *Corrosion Engineering, Science and Technology*, 43(2), pp. 186–191, 2008. DOI: 10.1179/174327807x214770. 62, 64

[183] L. Vrsalović, S. Gudić, and M. Kliškić, Salvia officinalis L. honey as corrosion inhibitor for CuNiFe alloy in sodium chloride solution. 2012. 62, 64

[184] I. Zaafarany, Corrosion inhibition of aluminum in aqueous alkaline solutions by alginate and pectate water-soluble natural polymer anionic polyelectrolytes. *Portugaliae Electrochimica Acta*, 30(6), pp. 419–426, 2012. DOI: 10.4152/pea.201206419. 62, 64

[185] E. Kusmierek and E. Chrzescijanska, Tannic acid as corrosion inhibitor for metals and alloys. *Materials and Corrosion*, 66(2), pp. 169–174, 2015. DOI: 10.1002/maco.201307277. 63, 64

Authors' Biographies

SHIMA GHANAVATI NASAB

Shima Ghanavati Nasab received her B.Sc. in applied chemistry in 2011 and her M.Sc. in analytical chemistry in 2014 from Payam Noor University. She joined the Shahrekord University as a Ph.D. student in the field of analytical chemistry and chemometrics and completed her research in 2018. During her research, she worked on pollutants removal and application of chemometric methods for the optimization. In 2017, she worked with Prof. Federico Marini group in Italy for nine months as a visiting Ph.D. student and finished two projects in QSPR in Lubricants and Classification of Honey samples. She has participated in Chemometrics school in Copenhagen University for two months under supervision of sophisticated teachers in the field of chemometrics. She was as a Talented student in M.Sc. and also introduced as a Top researcher in her Ph.D. period.

MEHDI JAVAHERAN YAZD

Mehdi Javaheran Yazd received his Bachelor's degree in mechanical engineering in 2008 with a research on heat treatment and HVOF from the IAU. He holds short courses in the science of materials and properties of metals. In 2019, he obtained a Master's degree in mechanical engineering with an energy conversion trend from the IAU. Since 2016, he is a member of the Young Researchers and Elite Club of the IAU. He began his specialty in energy field in 2009 and continued his research in the field of steam power plants. Since 2013 he has been focusing on energy optimization in industries, power plants, and buildings, and has followed energy modeling and exergy analysis.

ABOLFAZL SEMNANI

Abolfazl Semnani received his B.Sc. in chemistry from University of Isfahan in 1989. In 1991, he earned a M.Sc. and in 1995, got his Ph.D. in analytical chemistry from Shiraz University. His research interests are spectrophotometric and electrochemical study of crown ethers, conservation of cultural & historical works, environmental chemistry, industrial lubricants and chemometrics. He has received awards as a distinguished researcher from Shahrekord University in 1999, 2001, 2005, 2006, and 2008. He invented the process of synthesis of an additive for preparation of multigrade base oils from mineral base oils (Registration no. 388080661 in Iran). He has written more than 100 ISI papers.

HOMA KAHKESH

Homa Kahkesh graduated with a B.Sc. from Payam Noor University of Khorramshar in 2012. She earned an M.Sc. degree in analytical chemistry from Payam Noor university of Esfahan in 2014. She was admitted to the Shahid Chamran University of Ahvaz as a Ph.D. student in analytical chemistry in 2018. After receiving M.Sc. degree, she started to teach at the university. She also cooperated in writing some papers about the removal and corrosion field under the supervision of Professor Semnani. Her current interests are electrochemistry, corrosion, spectroscopy, and chemometrics.

NAVID RABIEE

Navid Rabiee graduated with an M.Sc. degree in inorganic chemistry from Shahid Beheshti University, Tehran, Iran, in 2018. During his M.Sc. research, he worked on porphyrin-based biosensors and the application of porphyrins and cobalt complexes on Dye Sensitized Solar Cells (DSSC) under the supervision of Prof. Nasser Safari. At the same time, he was focused on drug delivery systems based on biocompatible and biodegradable polymers associated with different types of sensitizers, especially porphyrins, under the supervision of Prof. Mohammad Rabiee at Amirkabir University of Technology, Tehran, Iran. In 2017, he joined ANNRG to collaborate with Prof. Mahdi Karimi's Research Lab at Iran University of Medical Science, Tehran, Iran, in association with Prof. Michael R. Hamblin from Harvard Medical University, Boston, MA, working on smart microcarriers/nanocarriers applied in therapeutic agent delivery systems employed for diagnosis and therapy of various diseases and disorders such as different cancers and malignancies, inflammations, infections, etc. Currently, he is a Ph.D. candidate in inorganic chemistry at Sharif University of Technology, Tehran, Iran, and working on different types of smart nanocarriers in collaboration with Amirkabir University of Technology and Iran University of Medical Sciences. Since 2018, he has participated in a collaboration with the Global Burden of Disease (GBD) project at the University of Washington, Seattle, Washington, in association with the Institute for Health Metrics and Evaluation (IHME), and has worked on some brilliant projects that have been published in *The Lancet*, *Nature*, and other highly-cited journals. His work has resulted in the publication of over 40 peer-reviewed journal articles, 5 books, and 3 conference papers.

MOHAMMAD RABIEE

Mohammad Rabiee, Ph.D., is an associate professor in the Biomedical Engineering Department at Amirkabir University of Technology, Tehran, Iran. His current research interests include smart drug delivery systems, tissue engineering, and biological sensors. He has published over 100 ISI papers and also over 70 international conference papers. In addition, he has been teaching and researching for 27 years at Amirkabir University of Technology, Tehran, Iran.

MOJTABA BAGHERZADEH

Mojtaba Bagherzadeh, Ph.D., is a professor in the Department of Chemistry at Sharif University of Technology, Tehran, Iran. His current research interests include inorganic chemistry, inorganic catalysis, and bioinorganic chemistry. He has published over 100 ISI papers, with the H-Index of 31.

Printed in the United States
by Baker & Taylor Publisher Services